Materials and surface engineering

Woodhead Publishing Reviews:

Mechanical Engineering

Series Editor: Professor J. Paulo Davim,
Department of Mechanical Engineering, University of Aveiro,
Portugal and Head of MACTRIB – Machining and Tribology Research Group
(email: pdavim@ua.pt)

Woodhead Publishing is pleased to publish this major Series of books entitled *Woodhead Publishing Reviews: Mechanical Engineering*. The Series Editor is Professor J. Paulo Davim, Department of Mechanical Engineering, University of Aveiro, Portugal and Head of MACTRIB – Machining and Tribology Research Group.

This research Series publishes refereed, high quality articles with a special emphasis on research and development in mechanical engineering from a number of perspectives including (but not limited to): machining and machine tools; tribology and surface engineering; materials and manufacturing processes; solid mechanics and structural mechanics; computational mechanics and optimization; mechatronics and robotics; fluid mechanics and heat transfer; renewable energies; biomechanics; micro- and nano-mechanics, etc. We seek authors, editors and contributors from a broad range of areas within the mechanical engineering discipline.

This Series examines current practises and possible future developments within the research field and industry-at-large. It is aimed at an international market of academics, practitioners and professionals working in the area. The books have been specially commissioned from leading authors, with the objective of providing the reader with an authoritative view of current thinking.

New authors: we would be delighted to hear from you if you have an idea for a book. We are interested in both shorter, practically orientated publications (45,000+ words) and longer, theoretical monographs (75,000–100,000 words). Our books can be single, joint or multi-author volumes. If you have an idea for a book, please contact the publishers or Professor J. Paulo Davim, the Series Editor.

Dr Glyn Jones
Woodhead Publishing Limited
Email: gjones@chandospublishing.com
www.woodheadpublishing.com

Professor J. Paulo Davim
Department of Mechanical Engineering,
University of Aveiro, Portugal
Email: pdavim@ua.pt
http://www2.mec.ua.pt/machining/pers-davim.htm

Woodhead Publishing Limited: established in 1989, Woodhead Publishing is a leading independent international publisher, publishing in the following main areas: food science, technology and nutrition; materials engineering; welding and metallurgy; textile technology; environmental technology; finance, commodities and investment; and mathematics.

Our ambitious publishing plans for the future will continue to bring you a range of authoritative reference books, professional texts and monographs, all written and produced to the exacting standards that have made Woodhead Publishing one of the UK's fastest growing independent publishers. All of our books are written in direct response to customers' needs by a truly international team of authors, ensuring they are designed for and relevant to a global audience.

Woodhead Publishing books are available worldwide, either direct, via our website, www.woodheadpublishing.com, or through booksellers and an international network of agents and representatives.

Professor J. Paulo Davim received his PhD in Mechanical Engineering from the University of Porto in 1997 and the Aggregation from the University of Coimbra in 2005. Currently, he is Aggregate Professor in the Department of Mechanical Engineering of the University of Aveiro and Head of MACTRIB - Machining and Tribology Research Group. He has more than 25 years of teaching and research experience in manufacturing, materials and mechanical engineering with special emphasis in machining and tribology. He is the Editor of four international journals, and also guest editor, editorial board member, reviewer and scientific advisor for many international journals and conferences. He has also published, as author and co-author, more than 30 book chapters and 300 articles in ISI journals (h-index 17) and conferences.

Bulk orders: some organisations buy a number of copies of our books. If you are interested in doing this, we would be pleased to discuss a discount. Please email wp@woodheadpublishing.com or telephone +44 (0) 1223 499140.

Related titles:

Mechatronics and manufacturing engineering: Research and development
Number 1 in the Woodhead Publishing Reviews: Mechanical Engineering
(ISBN 978-0-857-09-150-5)
Mechatronics is the blending of mechanical, electronic, and computer engineering into an integrated design. The main objective of this interdisciplinary engineering field is the study of automata from an engineering perspective, thinking on the design of products and manufacturing processes and systems. This book, the first in the **Woodhead Publishing Reviews: Mechanical Engineering Series**, will publish high quality articles (full research articles, review articles and cases studies) with a special emphasis on research and development in mechatronics and manufacturing engineering. Today, mechatronics has a significant and increasing impact on engineering with emphasis on the design, development and operation of manufacturing engineering systems. Mechatronics and manufacturing systems are well established and executed within a great number of industries including aircraft, automotive and aerospace industries; machine tools, moulds and dies product manufacturing, computers, electronics, semiconductor and communications, and biomedical.

Machining and machine-tools: Research and development
Number 3 in the Woodhead Publishing Reviews: Mechanical Engineering
(ISBN 978-0-85709-154-3)
The third book in the **Woodhead Publishing Reviews: Mechanical Engineering Series** includes high quality papers with a special emphasis on research and development in machining and machine-tools. Machining and machine tools is an important subject with application in several industries. Parts manufactured by other processes often require further operations before the product is ready for application. Traditional machining is the broad term used to describe removal of material from a workpiece, and covers chip formation operations including: turning, milling, drilling and grinding. Recently the industrial utilization of non-traditional machining processes such as electrical discharge machining, laser-beam machining, abrasive water jet machining and USM ultrasonic machining has increased. The performance characteristics of machine tools and the significant development of existing and new processes and machines, are considered. Nowadays, in Europe, USA, Japan and countries with emerging economies machine tools is a sector undergoing great technological development.

Tribology of natural fiber polymer composites
(ISBN 978-1-84569-393-0)
Tribology of natural fiber polymer composites examines the availability and processing of natural fiber composites and their structural, thermal, mechanical and tribological properties. It explores sources of natural fibers, their extractions and surface modification as well as properties of chemically modified natural fibers. It provides an overview of the tribology of polymer composites and the role of fiber reinforcement and filters in modifying tribological composites.

Details of these and other Woodhead Publishing books can be obtained by:

- visiting our web site at www.woodheadpublishing.com
- contact Customer Services (e-mail: sales@woodheadpublishing.com; fax: +44(0) 1223 832819; tel: +44(0) 1223 499140; address: Woodhead Publishing Limited, 80 High Street, Sawston, Cambridge CB22 3HJ, UK)

If you would like to receive information on forthcoming titles, please send your address details to Customer Services, at the address above. Please confirm which subject areas you are interested in.

Materials and surface engineering

Research and development

EDITED BY
J. PAULO DAVIM

WOODHEAD PUBLISHING

Oxford Cambridge Philadelphia New Delhi

Published by Woodhead Publishing Limited, 80 High Street, Sawston,
Cambridge CB22 3HJ, UK
www.woodheadpublishing.com

Woodhead Publishing, 1518 Walnut Street, Suite 1100, Philadelphia, PA 19102–3406, USA

Woodhead Publishing India Private Limited, G-2, Vardaan House, 7/28 Ansari Road,
Daryaganj, New Delhi – 110002, India
www.woodheadpublishingindia.com

First published 2012, Woodhead Publishing Limited

British Library Cataloguing in Publication Data
A catalogue record for this book is available from the British Library.

Library of Congress Control Number: 2011942113

Woodhead Publishing ISBN: 978-0-85709-151-2 (print)
ISBN: 978-0-85709-603-6 (online)

Typeset by RefineCatch Limited, Bungay, Suffolk
Printed in the UK and USA

Contents

List of figures

List of tables

Preface

Materials science is an interdisciplinary field involving the micro and nanostructure, processing, properties of materials and their application to various areas of engineering, technology and industry. Surface engineering is a truly interdisciplinary topic in materials science that deals with the surface of solid matter. Surface engineering techniques are being used in the automotive, aircraft, aerospace, missile, electronic, biomedical, textile, petrochemical, chemical, moulds and dies, machine tools and construction industries.

This research book aims to provide information on materials and surface engineering for modern industry. The initial chapter of the book provides a study of the scratch resistance of polymethylmethacrylate single-walled carbon nanotubes nanocomposites (the effect of modification by a room temperature ionic liquid). The second chapter is dedicated to local mechanical properties and nanoscale deformation of engineering metals and alloys. Chapter 3 presents an analysis of large-strain microindentation of crystalline polymers. Chapter 4 covers nanocomposite coatings (a review). Chapter 5 is dedicated to thermal sprayed WC–Co coatings for tribological application. Chapter 6 contains information on tribo performance of electroless Ni-P coatings. Chapter 7 covers tribological responses of materials with varying hardness and crystal structure during sliding on various surface textures. The last chapter of the book is dedicated to laser straight cutting of Ti-6Al-4V alloy (temperature and stress fields).

This book can be used as a research book for the final undergraduate engineering course or as a topic on materials and surface engineering at the postgraduate level. It can also serve as a useful reference for academics, materials and surface engineering researchers, mechanical, materials, manufacturing, physics and chemical engineers and professionals in surface engineering and related industries. The scientific interest in this book will be evident for many important centres of research including laboratories and universities throughout the world. Therefore, it is hoped this book will inspire and enthuse other researchers in this field.

The Editor acknowledges Woodhead Publishing for this opportunity and for their enthusiastic and professional support. Finally, I would like to thank all the chapter authors for their availability for this work.

J. Paulo Davim
University of Aveiro, Portugal
June 2011

About the contributors

Editor

J. Paulo Davim received his PhD in mechanical engineering from the University of Porto in 1997 and the Aggregation from the University of Coimbra in 2005. He is at present Aggregate professor in the Department of Mechanical Engineering of the University of Aveiro and Head of MACTRIB – Machining and Tribology Research Group. He has more than 25 years of teaching and research experience in manufacturing, materials and mechanical engineering with special emphasis on machining and tribology. He is the editor of nine international journals, guest editor, editorial board member, reviewer and scientific advisor for many international journals and conferences. He has also published more than 30 book chapters and 300 articles as author and co-author in refereed international journals (more than 150 in ISI Web Science, h-index=17) and conferences.

Authors

Syed Sohail Akhtar received his BSc and MSc degrees in Mechanical Engineering from NWFP University of Engineering and Technology, Pakistan in 1996 and 2000, respectively. He then received his PhD degree in Mechanical Engineering from King Fahd University of Petroleum and

Minerals (KFUPM), Saudi Arabia, in 2008, followed by a two-year post-doctorate fellowship in the same university. He has published more than 25 technical papers in refereed journals. He is currently working as Assistant Professor in the Mechanical Engineering department at KFUPM. Dr Akhtar's research areas are surface engineering, laser/thermal material processing and applications, materials characterization using modern techniques and applications of finite element analysis in multi-physics problems.

Alejandro Arribas has obtained his PhD degree in the research lines of the Materials Science and Engineering research group in the Materials and Manufacturing Engineering Department at the Technical University of Cartagena.

María-Dolores Bermúdez is head of the Materials Science and Engineering research group in the Materials and Manufacturing Engineering Department at the Technical University of Cartagena. Current research lines are: Tribology and Surface Science using room-temperature ionic liquids as lubricants under severe contact conditions, particularly in light alloys lubrication and extreme temperatures, and the development of new nanofluids by interaction of ionic liquids with nanophases and their use both as lubricants and as anti-wear polymer additives to develop new nanocomposites with enhanced tribological performance and scratch resistance.

Francisco-José Carrión is a senior researcher of the Materials Science and Engineering research group in the Materials and Manufacturing Engineering Department at the Technical University of Cartagena. His main research interests are focused in the study of structural, thermal, mechanical and tribological properties of polymers, polymer-matrix composites and nanocomposites, and the relationship between manufacturing processes and performance.

Constantinos A. Charitidis is an Associate Professor in the School of Chemical Engineering of the National Technical University of Athens. Currently, he is the Director of the Materials Science and Engineering Department and he is coordinating several projects funded by the Greek Ministry of Education and the European Commission. His main research interests are centred on the study of structural and nanomechanical properties of bulk thin-film multifunctional materials, tribology, contact mechanics, adhesion, fracture and fatigue of engineering materials, modelling of mechanical behaviour, nanomaterials processing, thin film technology, ion-beam and plasma assisted surface modification techniques, surface and interface properties, social and safety implications of nanotechnology. He has gained extensive R&D experience through collaborations with international research centres. He is a referee in international scientific journals, evaluator/scientific advisor of R&D projects, participant in EU and Greek funded R&D projects. He has published 170 peer-reviewed articles in scientific journals, has contributed 2 chapters in books and has edited 2 books.

Cayetano Espejo is developing a doctoral thesis in the research lines of the Materials Science and Engineering research group in the Materials and Manufacturing Engineering Department at the Technical University of Cartagena.

Satish V. Kailas is the Professor at the Department of Mechanical Engineering, Indian Institute of Science, Bangalore, India. He has around 15 years experience in the field of tribology. He did his undergraduate and graduate degrees and PhD in Mechanical Engineering. He worked as a guest researcher at NIST, USA before joining the Indian Institute of Science in 1997 as a faculty head. He is working on various fundamental aspects of tribology and friction stir welding and has over 100 publications. He is on the executive committee of the Tribology Society of India and on the editorial board of *Wear* journal.

Omer Keles received his BSc, MSc and PhD degrees in Mechanical Engineering from Gazi University in Turkey in 1998. He has published more than 40 technical papers in refereed international journals. He is currently working as Associate Professor in the Mechanical Engineering Department at Gazi University. His research areas are surface engineering, laser/thermal material processing and applications, and process control.

Kishore received his Bachelors Degree in both the Science and Engineering disciplines. A Masters in Engineering and Doctorate in the area of Physical Metallurgy followed this. He has authored over 200 papers in the areas of Structure Property Correlations, Tribology and Polymer Composites. His current interests include the study of surface texture in tribological behaviour, and utility of scanning electron microscopy for the study of failures in composites. Thermoset systems involving epoxies have been used as the matrix material in which polymeric fillers, as well as inorganic materials in the form of fibres, are introduced to characterize the mechanical properties.

E. Martínez-López has obtained his BEng degree in the research lines of the Materials Science and Engineering research group in the Materials and Manufacturing Engineering Department at the Technical University of Cartagena. Mr Martinez is currently in charge of a congeneration plant.

Michael R. Lovell is the interim chancellor of the University of Wisconsin-Milwaukee, Wisconsin, USA. He received his PhD in Mechanical Engineering from the University of Pittsburgh in 1994. His major accomplishments are the National Science Foundation CAREER award (1997), the Society of Manufacturing Engineers (SME) Outstanding Young Manufacturing Engineer Award (1999), the Outstanding International Publication on Bearings (1998), the American Society of Mechanical Engineers (ASME) Burt Newkirk Award in Tribology (2005), the Olympus Emerging Academic Innovator Award (2006) and he was named a fellow of ASME (2008). He is an Associate Editor of the ASME's *Journal of Tribology*. Dr. Lovell's primary research interests are in the areas of numerical simulation, novel manufacturing processes, tribology and micro and nanotechnology.

Y. Charles Lu is an Assistant Professor of Mechanical Engineering at the University of Kentucky. His research interests are centered in the general area of micromechanics and nanomechanics of polymers, elastomers, composites and advanced materials. Dr Lu has authored over 80 papers in referred journals and conference proceedings and over 100 technical reports. He was the recipient of the 2010 SAE International Ralph R. Teetor Educational Award, the ASEE Air Force Summer Faculty Fellowship Awards (2008, 2009), the Outstanding Mechanical Engineering Faculty Award (2009) and the Paducah Outstanding Professor Award (2009). Dr Lu received his PhD in Engineering Sciences from the University of Western Ontario in 2000 and is a licensed professional engineer in the state of Kentucky.

Pradeep L. Menezes received his PhD from the Department of Materials Engineering, Indian Institute of Science, Bangalore, India in 2008. Presently, he is working as a

Postdoctoral Research Associate in the Department of Industrial Engineering at the University of Wisconsin-Milwaukee, USA. He has authored over 100 research articles published in various international journals and conferences. He is a member of Society of Tribologists and Lubrication Engineers (STLE), reviewer of various tribology and materials engineering related journals. His current research interests include green and bio-manufacturing, energy efficiency and conservation, micro and nanoscale technologies, surface texturing, self-organization, metrology, tribology of green lubricants, green friction products and bio-composites.

R.M. Mohanty has been working as senior scientist at the Non-Ferrous Materials Technology Development Centre in India. His areas of research are extreme temperature materials engineering, thermal protection systems and hot structures, plasma technologies, thermal spray engineering, system engineering and research, development and production management. He is an alumnus of the Indian Institute of Technology Madras; the Institute of Technology, Varanasi; the National Institute of Technology Rourkela; and the Administrative Staff College of India. His rich experience in materials technologies, components and system engineering enables him to frame technological concepts, translating 'concept to hardware to system' in an efficient manner. He has numcrous national and international publications along with a United States patent in high temperature materials engineering.

Manish Roy graduated from the Indian Institute of Technology (Kharagpur) in 1985. Presently he is working as senior scientist in the Defence Metallurgical Research Laboratory, India. He is recipient of the 'Bilateral Exchange

Fellowship' of the Royal Society, London; the 'Lise Meitner' Fellowship of the Austrian Government; and a 'Marie Curie' Fellowship of the European Commission under an individual driven programme. His area of interest is materials and surface engineering for tribological application and high strain rate deformation. He has around 70 publications in various peer reviewed journals and some 15 presentations in various international events. He is a member of the editorial boards of several international journals.

Prasanta Sahoo is a Professor in the Department of Mechanical Engineering, Jadavpur University, Kolkata, India. His main research interests include tribology of materials and structural mechanics. He has authored a textbook on Engineering Tribology and a number of book-chapters. He has co-authored more than 200 technical papers. He is the Associate Editor of two international journals, member of the editorial boards of five international journals and on the review boards of over 30 international journals.

José Sanes is a senior researcher of the Materials Science and Engineering research group in the Materials and Manufacturing Engineering Department at the Technical University of Cartagena. He is currently studying real-time surface interactions and tribochemical processes in sliding contacts and optimizing the dispersions of nanophases in both thermosetting and thermoplastic matrices.

Loredana Santo graduated with honours in Aeronautic Engineering in 1994. From 1999 to 2005 she was researcher in Technologies and Manufacturing Systems in the Department of Mechanical Engineering of the University of Rome 'Tor Vergata'. She has been an Associate Professor since 2005. She is member of scientific committees of international conferences, and a member of editorial boards of international journals. She has been coordinator and co-organizer of the ESAFORM mini symposium 'Non-Conventional Processes' since 2007. Her scientific activity has been mainly focused on non-conventional processes, micromachining, innovative materials and processes. She is the author of 95 scientific papers published in international journals, proceedings of international conferences and books.

Bekir Sami Yilbas obtained his BSc, MSc and PhD degrees in Mechanical Engineering from Birmingham University in the UK. He worked at Birmingham, Glasgow and Erciyes Universities before joining King Fahd University of Petroleum and Minerals. He is currently a distinguished Professor there. His research area covers thermo-fluids, in particular thermoelectric laser machining applications and surface engineering. He has published over 475 papers in international journals (ISI) and presented 85 papers at conferences. He has been and is a serving editorial board member of seven international journals. He has received many awards for his scientific achievements, including the President of India's Prize for 1988; the Best Researcher awards from KFUPM in 1997, 2002, 2007; the Silver Jubilee Medal 2005 from the Silesian University of Technology, Poland; Doctor of Engineering from Birmingham University (2005); Donald Julius Groen Prize for 2007 from the Institute of Mechanical Engineers (IMechE), Manufacturing Industries Division, UK; and the Professor W. Johnson International Gold Medal for 2008.

1

Scratch resistance of modified polymethylmethacrylate nanocomposites

A. Arribas, M. D. Bermúdez, F. Carrión, C. Espejo, E. Martínez-López and J. Sanes, Universidad Politécnica de Cartagena, Spain

Abstract: Single-walled carbon nanotubes (NT) and single-walled carbon nanotubes modified (NTm) by the room-temperature ionic liquid 1-octyl, 3-methylimidazolium tetrafluoroborate ([OMIM]BF_4), were added in a 1wt.% to polymethylmethacrylate (PMMA) and processed by extrusion and injection moulding to obtain PMMA+NT and PMMA+NTm, respectively. Friction coefficients and abrasive wear of PMMA and PMMA nanocomposites from penetration depth, residual depth and viscoelastic recovery were determined under multiple scratching as a function of the number of passes of the indenter on the same groove. The effects of the moulding process, the additives and the sliding direction on the scratch resistance of PMMA were studied.

Key words: polymethylmethacrylate, carbon nanotubes, ionic liquids, scratch resistance.

1.1 Introduction

The first polymer nanocomposite containing carbon nanotube fillers was reported by Ajayan et al. (Ajayan, 2009). Since then, polymer nanocomposites have attracted increasing attention, particularly for their thermal, mechanical and electrical properties (Ajayan, 2000; Spitalski, 2010; Moniruzzaman, 2006; Ramanathan, 2005; Thostenson, 2001; Xie, 2005; Wang, W., 2008). There are fewer studies focused on the friction, wear and scratch behaviour of these materials, however, although there exists a growing interest in the development of polymers and polymer nanocomposites with enhanced surface stability and tribological performance (Burris, 2007; Brostow, 2008; Huang, 2009; Dasari, 2009; Pelletier, 2010; Dong, 2005; Cai, 2004; Dong, 2008; Wang, C., 2008; Giraldo, 2008). In a very recent review, A. Dasari et al. (2009) concluded that significant contributions are needed to the fundamental tribological knowledge of polymer/carbon nanotube nanocomposites.

The atomically smooth surface without dangling atoms, and the durability of the lattice structure of carbon nanotubes, makes them a good solid lubricant with a reported ultra-low friction coefficient of 0.01 against diamond (Mylvaganam, 2009). However, the friction coefficient values for carbon nanotubes can vary from 0.01 to 2.0. Moreover, some results on the tribological behaviour of carbon nanotube-reinforced polymer composites have even led to contradictory conclusions. In some cases, the addition of carbon nanotubes into polymer matrices reduces both friction and wear, but in other cases they do not improve the tribological performance of the polymer.

Anisotropic response is another aspect to be taken into account. The friction coefficient of films of aligned single-walled carbon nanotubes against a diamond tip (Mylvaganam, 2009) exhibits extremely low friction when sliding is conducted on horizontally oriented carbon nanotubes.

Room temperature ionic liquids have shown outstanding lubricating performance for metals, polymers and ceramic materials under severe conditions (Jiménez, 2006; Sanes, 2006; Sanes, 2007; Carrión, 2007; Bermúdez, 2009; Minami, 2009; Zhou, 2009; Torimoto, 2010; Palacio, 2010). The combination of ionic liquids and nanophases has given rise to new nanofluids which are finding some promising applications (Fukushima, 2003, 2006 and 2007; Wang, J., 2008; Wang, B., 2010; Carrión, 2011). In particular, the ability of ionic liquids to disperse and functionalize carbon nanotubes has come to be of great interest. The interfacial strength between carbon nanotubes and a polymer matrix increases dramatically when the carbon nanotube surface is chemically modified (Barber, 2006).

Imidazolium ion-based room-temperature ionic liquids grounded with pristine single-walled carbon nanotubes form gels by physical cross-linking of the nanotube bundles, mediated by local molecular ordering of the ionic liquids (Fukushima, 2003). The combination of a room-temperature ionic liquid and single-walled carbon nanotubes has been recently used as lubricant of polycarbonate/stainless steel contact (Carrión, 2011).

In an immediately preceding study (Carrión, 2010), we have described the preparation of compression moulded nanocomposites and the effect of the addition of nanotubes (NT) and modified nanotubes (NTm) on friction and wear of thermoplastics under sliding conditions against steel in the pin-on-disc configuration. Under such conditions, a predominantly adhesive wear mechanism takes place. A significant wear reduction was achieved for the nanocomposites with respect to the base polymers. Polymethylmethacrylate with modified nanotubes (PMMA+NTm) presented a wear rate decrease of 58% with respect to PMMA+NT and of 63% with respect to PMMA. It was observed that for PMMA+NTm, the damage produced by the steel pin was limited to the outer surface layers. The good tribological performance of the PMMA nanocomposites was attributed to the uniform dispersion of the additives within the polymer matrix which inhibits crack propagation and fracture. In the present study, multiple scratch testing conditions have been used in order to determine the abrasive wear behaviour of the new nanocomposites.

1.2 Experimental data

Single-walled carbon nanotubes (1–2 nm diameter; 5–30 μ length) (NanoAmor Inc.) obtained by the catalytic Chemical Vapour Deposition method, with >90% content in single-walled carbon nanotubes (96.30% C; 0.08% Al; 0.41% Cl; 2.91% Co; 0.29% S), were used as-received (Fig. 1.1). Transmission electron microscopy (TEM) micrographs were obtained with a high resolution JEOL JEM 2100. The room-temperature ionic liquid 1-octyl, 3-methylimidazolium tetrafluoroborate ([OMIM]BF_4) (Fluka, Germany) (>98% purity) was used as-received (Fig. 1.2).

Ionic liquid-modified single-walled carbon nanotubes (NTm) (Fig. 1.3) were obtained after mechanical mixing and ultrasonic dispersion in [OMIM]BF_4 following the previously described procedure (Carrión, 2010).

After heating at 90°C for 16 hours, PMMA powders (Buehler; Germany), or their mixtures with a 1 wt.% proportion of NT or NTm, were milled in an ultra-centrifugal mill at 18,000 rpm for 30 s. Tensile test coupons (according to ASTM D-256) (Fig. 1.4) were extruded in a Leistritz ZSE18 HP extrusion machine at 250°C under a pressure

Figure 1.1 TEM micrograph of as-received single-walled carbon nanotubes (NT)

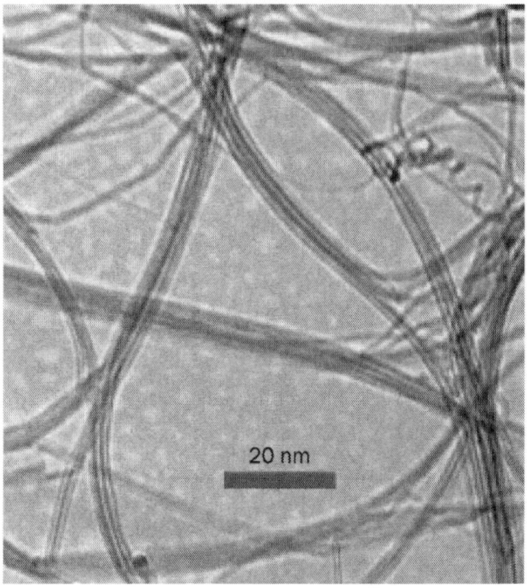

Figure 1.2 Chemical formula of the room-temperature ionic liquid [OMIM]BF_4

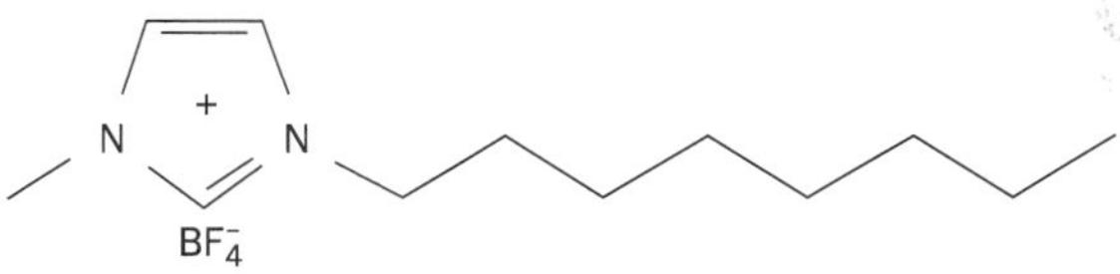

of 1 MPa and injection moulded at 225°C under 10 MPa in a DEU (Spain) 250H55 mini VP injection machine with a mould temperature of 55°C and a holding pressure of 7 MPa.

Figure 1.5 shows a transmission electron microscopy (TEM) micrograph of PMMA+NTm. Shore D hardness values (Table 1.1) were determined with a TH210 durometer, according to ASTM D2240 standard. PMMA and PMMA+NT show the same hardness values. Only the addition of NTm reduces hardness, probably due to the presence of the fluid ionic liquid phase.

Figure 1.3 TEM micrograph of single-walled carbon nanotubes modified (NTm) after treatment with the ionic liquid

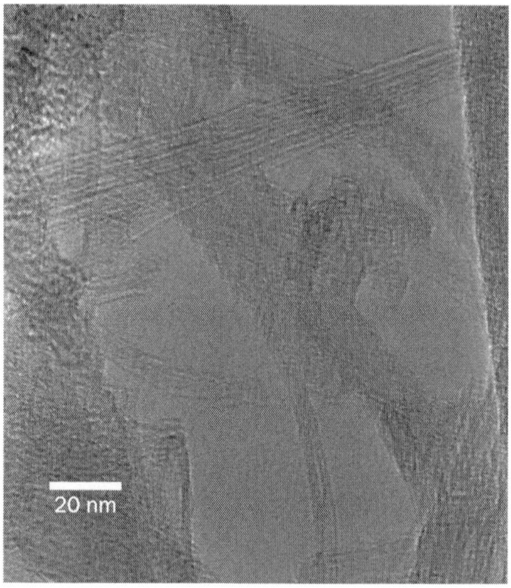

Figure 1.4 Test coupon for scratch tests. The arrow shows the sliding direction, parallel to injection flow

Table 1.1 Shore D hardness

PMMA	PMMA+NT	PMMA+NTm
85.8 (1.1)	85.5 (1.0)	82.7 (0.3)

Raman spectra (Fig. 1.6) were collected using a Nicolet Almega XR Raman spectrometer from Thermo Electron. A laser of 780 nm at 100% of power and 50 μm of aperture was used. As can be observed in Fig. 1.6, the PMMA+NTm

Figure 1.5 TEM micrograph of PMMA+NTm

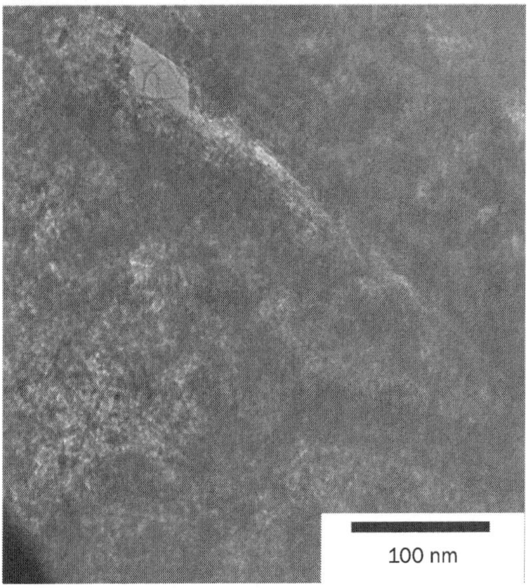

Figure 1.6 Raman spectra of PMMA, NTm and PMMA+NTm

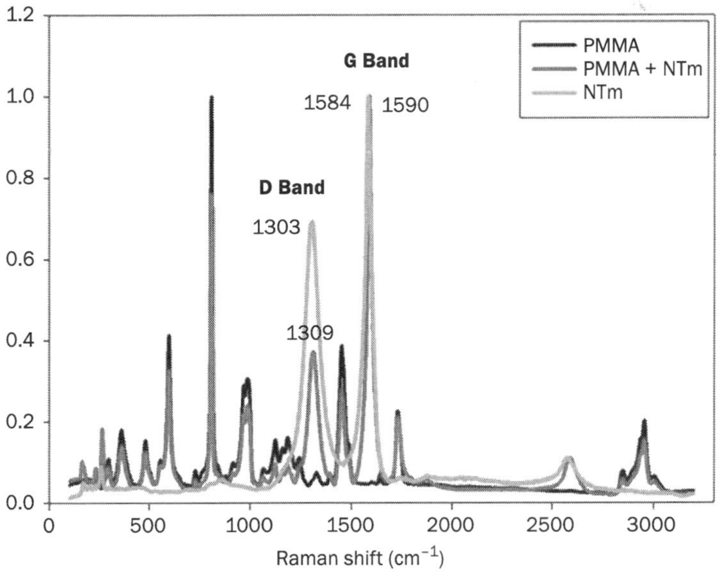

nanocomposite exhibits both the bands due to the base polymers, and the D and G bands characteristic of the nanotubes, with a lower intensity.

A differential scanning calorimeter (DSC) 822e (Mettler-Toledo) was used to determine glass transition temperatures (Tg) (Table 1.2). 15 mg samples were heated under an inert N_2 atmosphere (50 ml/min), at a heating rate of 10°C/min. The T_g values obtained for the injected samples (Table 1.2) are lower than those of compression moulded discs (Carrión, 2010), particularly in the case of the base polymer, PMMA, and the nanocomposite containing the nanotubes modified by previous treatment with the ionic liquid (PMMA+NTm). This decrease in T_g values for the injected samples could suggest a higher chain mobility due to preferential orientation in the injection flow direction than in the unoriented compression moulded samples.

Scratch tests according to ASTM D7027-05 (Fig 1.7) were carried out with a MTR 3/50–50/NI Microtest (Spain) Scratch Tester, with a diamond-tip indenter (200 μm diameter and 120° cone angle), using an inductive probe with a measurement range of ±250 μm and an accuracy of ±0.5 μm, following the previously described procedure (Bermúdez, *Polymer*, 2005; Bermúdez, *E-Polymers*, 2005; Sanes, 2010; Bermúdez, 2010).

In the multiple scratch tests described here, 15 scratches were successively performed on the same track, with a

Table 1.2 T_g values (DSC)

Material	T_g(°C)
PMMA	92.1
PMMA+NT	100.7
PMMA+NTm	91.7

Figure 1.7 Schema of scratch test configuration

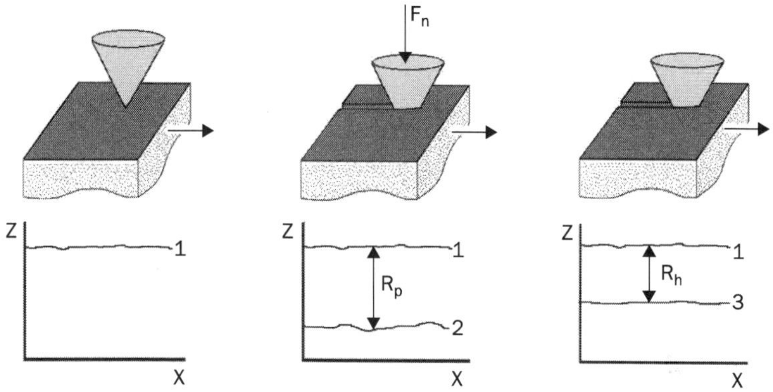

scratch length of 5 mm at a scratch velocity of 5 mm/min, under a normal load of 5 N. Each run included a pre-scan to establish the initial topology of the surface, a scratch to measure the penetration depth (R_p) under load, and a post-scan to measure the residual depth (R_h) after the viscoelastic recovery or healing has taken place (Sanes, 2010). The percentage of viscoelastic recovery (f) is calculated from mean R_p and R_h values, as:

$$[1 - (R_h/R_p)] \times 100 \qquad [1.1]$$

3-D surface topography images were obtained by means of a Talysurf CLI optical profiler. Scanning electron microscopy (SEM) images were obtained using a Hitachi S3500N. The samples were gold coated with a SC7640 Sputter Coater before observation.

1.3 Scratch resistance

Figure 1.8 shows the evolution of friction coefficients under multiple scratching. In this case, when sliding is parallel to

Figure 1.8 Friction coefficients as a function of the number of scratches

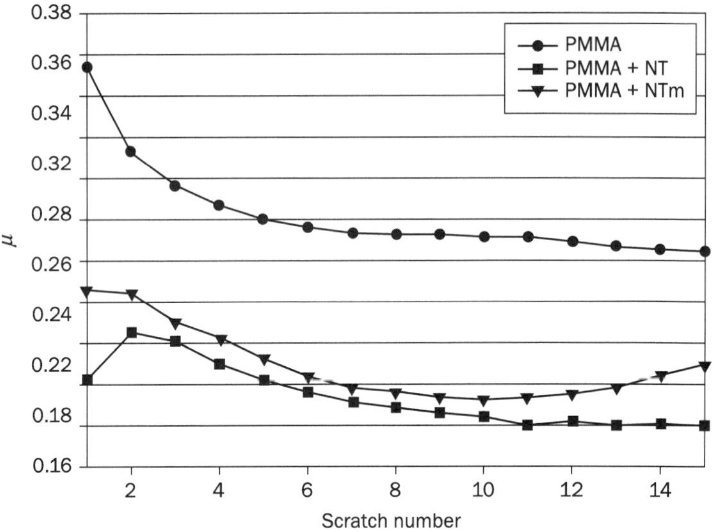

injection flow, both NT and NTm show an efficient internal lubricant behaviour with respect to the base polymer. The friction values for both nanocomposites are very similar, with a maximum friction reduction of 32% obtained for PMMA+NT with respect to PMMA.

Figure 1.9 shows the evolution of the instantaneous penetration depth of the diamond indenter with the number of successive scratches in the three materials, when the sliding is performed on the central region of the injected coupons (Fig. 1.4) along the direction parallel to the injection flow; that is, parallel to the orientation of the polymer chains and the nanotubes. Both nanocomposites show penetration depth values lower than those of the base polymer. The lowest penetration depth values and the lowest increases with each scratch are obtained for the nanocomposite containing single-walled carbon nanotubes. The scratch resistance is a measurement of the dynamic hardness of the

Figure 1.9 Instantaneous penetration depth as a function of the number of scratches

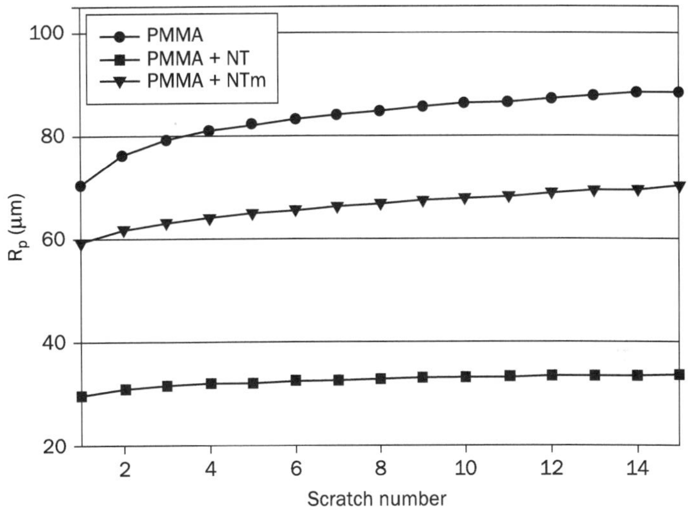

material. In this case, the highest static hardness value found for PMMA+NT (Table 1.1) is in agreement with its lower penetration depth.

Another important difference between the base polymer and the nanocomposites is the increase in penetration depth with the number of scratches. For the first 4 scratches, PMMA shows a strong increase in R_p, in contrast with the nanocomposites and particularly PMMA+NT, which shows an asymptotic behaviour from the first passes of the indenter.

Figure 1.10 shows the final permanent damage due to abrasive wear under multiscratching, after the viscoelastic recovery has taken place. The final surface damage on the base polymer is very severe, with an increase in residual depth of 77% after 15 scratches, with respect to PMMA+NT. Again, as observed for the instantaneous penetration, the order of scratch resistance is: PMMA+NT > PMMA+NTm > PMMA. However, while in instantaneous penetration the behaviour of PMMA+NTm is

Figure 1.10 Residual depth after viscoelastic recovery as a function of the number of scratches

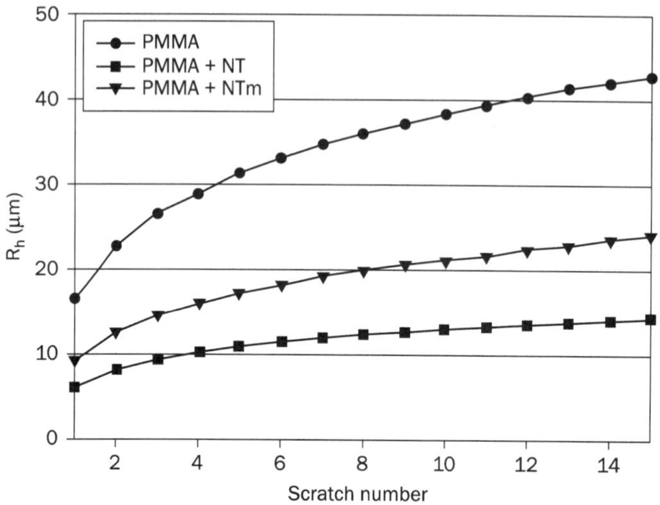

closer to the base polymer than to PMMA+NT, after the viscoelastic recovery the final damage on PMMA+NTm is closer to PMMA+NT (see Table 1.3), the material with the highest hardness (Table 1.1). This suggests that the nanocomposite containing nanotubes previously treated with ionic liquid have a greater capacity for viscoelastic recovery.

Figure 1.10 shows that the neat polymer presents higher increases in residual depth with each successive scratch than the nanocomposites, which show a more asymptotic behaviour, due to the strain hardening effect under multiple scratching which has been previously described for most thermoplastic materials (Bermúdez, 2005; Brostow, 2006 and 2008). Again, both nanocomposites reduce the permanent damage with respect to the base polymer, but the lowest residual depth is obtained for the material containing single-walled carbon nanotubes which have not been previously treated with ionic liquid.

Figure 1.11 Percentage of viscoelastic recovery as a function of the number of scratches

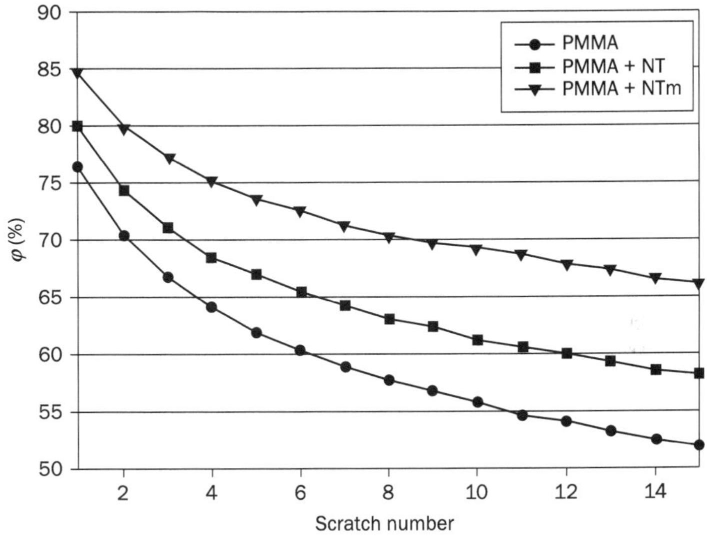

The percentages of viscoelastic recovery for each material have been calculated according to Eq. [1.1] and are represented in Fig. 1.11. In this case, the order of viscoelastic recovery ability is: PMMA+NTm > PMMA+NT > PMMA. The base polymer not only shows the lowest recovery values, but the difference with the nanocomposites increases as the number of scratches increases, thus the highest difference takes place after 15 scratches. Final values of scratch resistance and friction coefficients after 15 scratches are given in Table 1.3. This behaviour could be due to a higher polymer chain mobility in PMMA+NTm, in agreement with its lower glass transition temperature (Table 1.2) and to the alignment of the nanotubes parallel to flow and the scratching direction. This alignment and mobility could be responsible for the lower friction values (Fig. 1.8) found for the nanocomposites, with respect to the pure polymer.

Table 1.3 Scratch resistance after 15 scratches

	Scratch tests results			
Material	Penetration depth (R_p; μm)	Residual depth R_h; μm)	Viscoelastic recovery (φ; %)	Friction coefficient (μ)
PMMA	88.43	42.74	52	0.26
PMMA+NT	33.49	14.03	58	0.18
PMMA+NTm	70.05	23.81	66	0.21

1.4 Wear mechanisms

Figure 1.12 shows SEM micrographs of the grooves on the three materials after 15 scratches. Three different mechanisms are observed. PMMA shows a smooth, very wide and deep scar, without wear debris or cracks. PMMA+NT shows a very narrow groove with multiple wear debris particles, both along the edge and inside the groove. Finally, PMMA+NTm shows a plastic deformation mechanism with material accumulation of the edges, delamination and crack propagation in the direction perpendicular to the indenter sliding direction.

Figure 1.12 SEM micrographs of grooves after 15 scratches: a) PMMA; b) PMMA+NT; c) PMMA+NTm

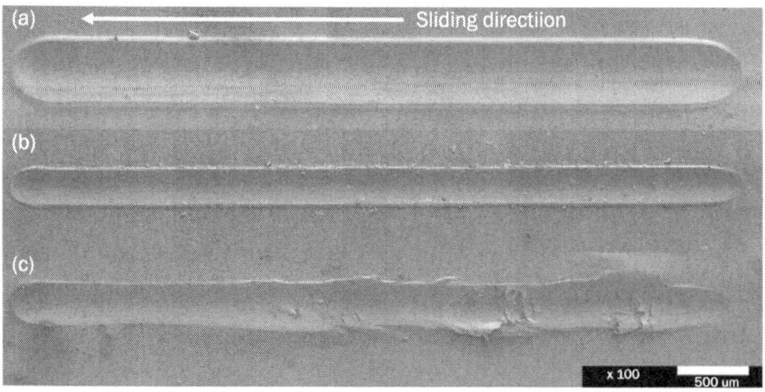

Figure 1.13 3-D surface profiles of the grooves after 15 scratches

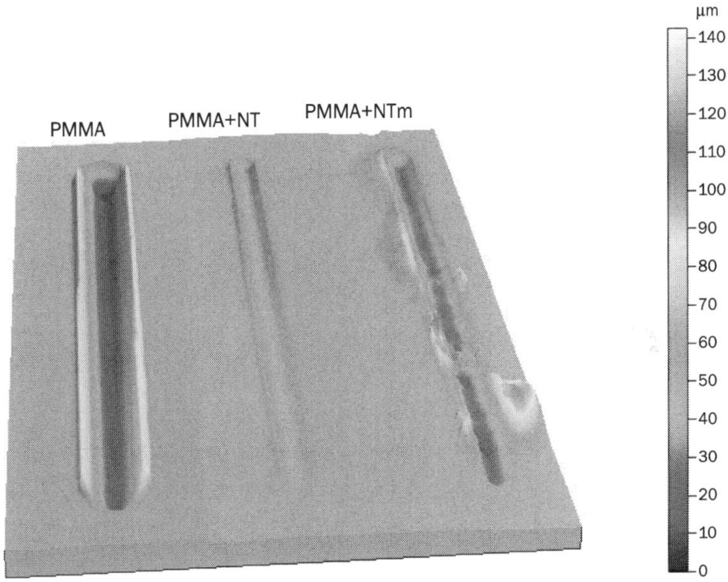

Figure 1.13 shows the 3-D surface profiles of the grooves after 15 scratches. This confirms the plastic deformation for PMMA, the presence of wear debris for PMMA+NT, and the crack propagation across the groove on PMMA+NTm. In contrast with the rest of the materials, the deformation mechanism for PMMA+NTm presents more severe plastic deformation and crack propagation on the first half (with respect to the indenter sliding direction (Fig. 1.12) of the groove. This non-uniform behaviour could be due to the presence of defects, which induce the nucleation of cracks.

1.5 Conclusions and future work

The addition of single-walled carbon nanotubes or single-walled carbon nanotubes modified by a room temperature ionic liquid reduces friction coefficients, instantaneous penetration depth and final residual depth and increases the percentages of viscoelastic recovery of injection moulded PMMA, under multiple scratching conditions with sliding parallel to injection flow. The highest instantaneous and residual depth reductions are found for PMMA+NT, containing pristine carbon nanotubes, which reduce chain mobility and plastic deformation of the polymer, but the highest viscoelastic recovery increase is found for PMMA+NTm, containing carbon nanotubes modified by ionic liquid, which increase chain mobility and plastic deformation.

The influence of the ionic liquid composition, the interaction with the nanotubes, the use of multi-walled carbon nanotubes instead of single-walled ones, the influence of the additives content, processing conditions and sliding direction with respect to the polymer chains and additives orientation, are future research lines to be followed with the objective of optimizing the performance of the new nanocomposites. The results described here show that carbon nanotube nanocomposites increase not only the adhesive wear resistance, but also the abrasive wear resistance of the polymers, and could be used in applications where this is the main surface damage mechanism, such as coatings.

1.6 Acknowledgements

The authors wish to acknowledge the financial support of MICINN (España), the European FEDER program (MAT2008–01670), and the 'Programa de Generación de Conocimiento Científico de Excelencia' Fundación Séneca (08596/PI/08).

1.7 References

Ajayan PM, Schadler LS, Giannaris C, Rubio A (2000), 'Single-walled carbon nanotube-polymer composites: strength and weakness', *Adv Mater*, **12**: 750–3.

Ajayan PM, Stephan O, Colliex C, Trauth D (2009), 'Aligned carbon nanotube arrays formed by cutting a polymer resin-nanotube composite', *Science*, **265**: 1212–4.

Barber AH, Cohen SR, Eitan A, Schadler LS, Wagner HD (2006), 'Fracture transitions at a carbon nanotube/polymer interface', *Adv Mater*, **18**: 83–7.

Bermúdez MD, Brostow W, Carrión-Vilches FJ, Cervantes JJ, Pietkievicz D (2005), 'Friction and multiple scratch behaviour of polymeronomer liquid crystal systems', *Polymer*, **46**: 347–62.

Bermúdez MD, Brostow W, Carrión-Vilches FJ, Sanes J (2010), 'Scratch resistance of polycarbonate containing ZnO nanoparticles: effects of sliding direction', *J Nanosci Nanotechnol*, **10**: 6683–9.

Bermúdez MD, Brostow W, Carrión FJ, Cervantes JJ (2005), 'Wear of thermoplastics determined by multiple scratching', *E-Polymers*, No. 001.

Bermúdez MD, Jiménez AE, Sanes J, Carrión FJ (2009), 'Ionic liquids as advanced lubricants', *Molecules*, **14**: 2888–908.

Brostow W (2008), 'Tribology of polymer-based materials', *Mater Res Innov*, **12**: 102–4.

Brostow W, Lobland HEH (2006), 'Sliding wear, viscoelasticity and brittleness of polymers', *J. Mater Res*, **21**: 2422–8.

Brostow W, Lobland HEH (2010), 'Brittleness of materials: implications for composites and a relation to impact strength', *J Mater Sci*, **45**: 242–50.

Burris DL, Boesl B, Bourne GR, Sawyer WG (2007), 'Polymeric materials for tribological applications', *Macromol Mater Eng*, **292**: 387–402.

Cai H, Yan FY, Xue QJ (2004), 'Investigation of tribological properties of polyimide/carbon nanotube nanocomposites', *Mater Sci Eng A-Struct Mater Prop Microstruct Process*, **364**: 94–100.

Carrión FJ, Espejo C, Sanes J, Bermúdez MD (2010), 'Single-walled carbon nanotubes modified by ionic liquid as antiwear additives of thermoplastics', *Compos Sci Technol*, **70**: 2160–7.

Carrión FJ, Sanes J, Bermúdez MD, Arribas A, Bermúdez MD (2011), 'New single-walled carbon nanotubes-ionic liquid lubricant. Application to polycarbonate-stainless steel sliding contact', *Tribol. Lett*, **41**: 199–207.

Carrión FJ, Sanes J, Bermúdez MD (2007), 'Effect of ionic liquids on the structure and tribological properties of polycarbonate-zinc oxide nanodispersion', *Mater Lett*, **61**: 4531–5.

Dasari A, Yu ZZ, Mai YW (2009), 'Fundamental aspects and recent progress on wear/scratch damage in polymer nanocomposites', *Mater Sci Eng R*, **63**: 31–80.

Dong B, Wang C, He BL, Li HL (2008), 'Preparation and tribological properties of poly(methylmethacrylate)/styrene/MWNTs copolymer nanocomposites', *J Appl Polym Sci*, **108**: 1675–9.

Dong B, Yang Z, Huang Y, Li HL (2005), 'Study on tribological properties of multi-walled carbon nanotubes/epoxy resin nanocomposites', *Tribol Lett*, **20**: 251–4.

Fukushima T, Aida T (2007), 'Ionic liquids form soft functional materials with carbon nanotubes', *Chem-Eur J*, **13**: 5048–58.

Fukushima T, Kosaka A, Ishimura Y, Yamamoto T, Takigawa T et al. (2003), 'Molecular ordering of organic molten salts triggered by single-walled carbon nanotubes', *Science*, **300**: 2072–4.

Fukushima T, Kosaka A, Yamamoto Y, Aimiya T, Notazawa S et al. (2006), 'Dramatic effect of dispersed carbon nanotubes on the mechanical and electroconductive properties of polymers derived from ionic liquids', *Small*, **2**: 554–60.

Giraldo LF, Brostow W, Devaux E, Lopez BL, Perez LD (2008), 'Scratch and wear resistance of polyamide 6 reinforced with multiwall carbon nanotubes', *J Nanosci Nanotechnol*, **8**: 3176–83.

Giraldo LF, Lopez BL, Brostow W (2009), 'Effect of the type of carbon nanotubes on tribological properties of polyamide 6', *Polym Eng Sci*, **49**: 896–902.

Huang YL, Yuen SM, Ma CM, Chuang CY, You KC et al. (2009), 'Morphological, electrical, electromagnetic interference (EMI) shielding and tribological properties of functionalized multi-walled carbon nanotube/ polymethylmethacrylate (PMMA) composites', *Compos Sci Technol*, **69**: 1991–6.

Jiménez AE, Bermúdez MD, Iglesias P, Carrión FJ, Martínez-Nicolás G (2006), '1-N-alkyl-3-methylimidazolium ionic liquids as neat lubricants and lubricant additives in steel-aluminium contacts', *Wear*, **260**: 766–82.

Minami I (2009), 'Ionic liquids in tribology', *Molecules*, **14**: 2262–9.

Moniruzzaman M, Winey, KI (2006), 'Polymer nanocomposites containing carbon nanotubes', *Macromolecules*, **39**: 5194–205.

Mylvaganam K, Zhang LC, Xiao KQ (2009), 'Origin of friction in films of horizontally oriented carbon nanotubes sliding against diamond', *Carbon*, **47**: 1693–700.

Palacio M, Bhushan B (2010), 'A review of ionic liquids for green molecular lubrication in nanotechnology', *Tribol Lett*, **40**: 247–68.

Pelletier H, Gauthier C, Schirrer R (2010), 'Wear simulation of polymer using multiscratch test procedure', *Tribol Lett*, **37**: 507–15.

Ramanathan T, Liu H, Brinson LC (2005), 'Functionalized SWNT/polymer nanocomposites for dramatic property improvement', *J Polym Sci B-Polym Phys*, **43**: 2269–79.

Sanes J, Carrión FJ, Bermúdez MD (2010), 'Effect of the addition of room temperature ionic liquid and ZnO nanoparticles on the wear and scratch resistance of epoxy resin', *Wear*, **268**: 1295–302.

Sanes J, Carrión FJ, Bermúdez MD, Martínez-Nicolás G (2006), 'Ionic liquids as lubricants of polystytrene and polyamide 6-steel contacts. Preparation and properties of new polymer-ionic liquid dispersions', *Tribol Lett*, **21**: 121–33.

Sanes J, Carrión FJ, Jiménez AE, Bermúdez MD (2007), 'Influence of temperature on PA 6-steel contacts in the presence of an ionic liquid lubricant', *Wear*, **263**: 658–62.

Spitalsky Z, Tasis D, Papagelis K, Galiotis C (2010), 'Carbon nanotube–polymer composites: Chemistry, processing, mechanical and electrical properties', *Prog Polym Sci*, **35**: 357–401.

Thostenson ET, Ren Z, Chou T (2001), 'Advances in the science and technology of carbon nanotubes and their composites: a review', *Compos Sci Technol*, **61**: 1899–912.

Torimoto T, Tsuda T, Okazaki K, Kuwawata S (2010), 'New frontiers in materials science opened by ionic liquid', *Adv Mater*, **22**: 1196–221.

Wang B, Wang X, Lou W, Hao J (2010), 'Rheological and tribological properties of ionic liquid-based nanofluids containing functionalized multi-walled carbon nanotubes', *J Phys Chem C*, **114**: 8749–54.

Wang C, Xue T, Dong B, Wang Z, Li HL (2008), Polystyrene-acrylonitrile-CNTs nanocomposites preparation and tribological behavior research', *Wear*, **265**: 1923–6.

Wang J, Chu H, Li Y (2008), 'Why single-walled carbon nanotubes can be dispersed in imidazolium-based ionic liquids', *ACS Nano*, **2**: 2540–6.

Wang W, Ciselli P, Kuznetsov E, Peijs T, Barber AH (2008), 'Effective reinforcement in carbon nanotube–polymer composites', *Phil Trans R Soc A*, **366**: 1613–26.

Xie XL, Mai YW, Zhou XP (2005), 'Dispersion and alignment of carbon nanotubes in polymer matrix: A review', *Mater Sci Eng R*, **49**: 89–112.

Zhou F, Liang Y, Liu W (2009), 'Ionic liquid lubricants: designed chemistry for engineering applications', *Chem Soc Rev*, **28**: 2590–9.

Nanomechanical properties and nanoscale deformation of engineering metals and alloys

C. Charitidis, National Technical University of Athens, Greece

Abstract: Metals and alloys are widely used in transport and industry applications, due to their functionality. The nanomechanical properties and nanoscale deformation of engineering metals and alloys are investigated, together with pile-up/sink-in deformation mechanism analysis. The study of the onset of dislocation plasticity and possible phase transformations can be performed during the nanoindentation test. A yield-type pop-in occurs upon low applied load, representing the start of phase transformation. Power-law function has been fitted to the loading data, to determine the displacement offset. Hardness at infinite displacement was extracted, and compared with bulk (~400 nm) hardness of the materials. It is shown that nanoindentation is a convenient measurement of local mechanical properties and plasticity of probed materials, including the study of fundamental materials physics and elementary mechanisms.

Key words: nanoindentation, onset of dislocation plasticity, engineering metals and alloys.

2.1 Introduction

Compressed or stretched loads may result in residual strain in metals and alloys. When the amount of such loads and the range of affected areas are very small, residual strains aroused by loads are not only very small but also appear in a very confined area, which presents difficulties for measurement (Wang et al., 2009). The strength of metals and alloys is strongly influenced by grain size (Kumar et al., 2003; Gleiter, 2000). Materials in the nanocrystalline regime are characterized by superior yield and fracture strength, improved wear resistance, superplasticity observed at relatively low temperatures and high strain rates as compared with their microcrystalline counterparts (Masumura et al., 1998; Jeong et al., 2001; Hayes et al., 2004; Jia et al., 2003) This has led to increased attention on understanding their mechanical properties and deformation mechanisms, summarized in recent review articles (Kumar et al., 2003; Suryanarayana, 2005). In fact, microstructures resulting in high long-life fatigue resistance will generally yield lower thresholds for fatigue crack growth, especially in ultrafine and nanocrystalline regimes (Cavaliere, 2009).

Conventionally instrumented nanoindentation equipment provides load–displacement curves for a monotonically increasing load allowing precise determination of different properties such as yield strength, hardness, hardening behaviour, wear characteristics, etc. The usefulness of nanoindentation obtaining the fundamental mechanical properties of materials has been widely demonstrated elsewhere (Cao and Lu, 2005; Bucaille et al., 2003; Dao et al., 2001; Cheng and Cheng, 2004; Doerner and Nix, 1998). Such a technique has much broader applications, ranging from an understanding of fundamental materials physics (Schuh, 2006) to its use as flexible mechanical probe (Schwaiger et al., 2003). The indentation hardness (mean pressure underneath the indenter) of a solid is commonly used as resistivity to plastic deformation. Indentation hardness is equal to the load applied in pressing a hard indenter of specified geometry into a specimen, divided by the area of the resulting impression. For the Vickers hardness, the area in question is the actual contact area between the indenter and the specimen (Elmustafa and Stone, 2003).

The nanoindentation test can provide information about the mechanical behaviour of the material when it is deformed at the sub-micron scale. The method developed by Oliver and Pharr (Oliver and Pharr, 1992; King, 1987) allows the hardness and elastic modulus to be determined from the nanoindentation load displacement data. However, it has been observed by several authors that hardness in nanoindentation tends to be overestimated due to a high size dependence (Hutchinson, 2000; Nix and Gao, 1998) of the tests (the so-called Indentation Size Effect (ISE)). ISE means that hardness at shallow displacements is much greater than at greater displacements (Tymiak et al., 2001). Therefore, on the nanoindentation range, the hardness value will be strongly dependent on the displacement of the indentation. Nevertheless, the ISE has not only been observed in nanoindentation tests but has also been detected for low load Vickers measurements (Miyahara et al., 2001).

Continuum mechanics-based analysis cannot explain ISE, since the hardness value should be independent of the applied load or, equivalently, of the indent size (Elmustafa and Stone, 2003). Several mechanisms that explain the ISE have been suggested, including the effect of the friction between the indenter and the sample (Li et al., 1993), the presence of a work-hardened surface layer (Pelletier et al., 2000), the lack of measurement capabilities at shallow indentations or the presence of surface layers, oxides and chemical contamination (Atkinson, 1991).

2.2 Experimental data

Nanoindentation testing was performed with a Hysitron TriboLab® nanomechanical test instrument, which allows the application of loads from 1 to 10 000 μN and records the displacement as a function of applied loads with a high load resolution (1 nN) and a high displacement resolution (0.04 nm). The TriboLab® employed in this study is equipped with a Scanning Probe Microscope (SPM), in which a sharp probe tip moves in a raster scan pattern across a sample surface using a three-axis piezo positioner (Fig. 2.1). In all nanoindentation tests a total of 10 indents are averaged to determine the mean hardness (H) and elastic modulus (E) values for statistical purposes, with a spacing of 50 μm, in a clean area environment with 45 per cent humidity and 23°C ambient temperature. In order to operate under closed loop load or displacement control, a feedback control option was used. All nanoindentation measurements were performed with the standard three-sided pyramidal Berkovich probe, with an average radius of curvature of about 100 nm (Charitidis, 2010a).

Based on the half-space elastic deformation theory, H and E values can be extracted from the experimental data (load–displacement curves) using the Oliver–Pharr method (Oliver and Pharr, 1992; King, 1987), where derived expressions for calculating the elastic modulus from indentation experiments

Figure 2.1 Instrumentation of nanoindentation testing of metals and alloys

are based on Sneddon's elastic contact theory (Sneddon, 1948):

$$E_r = \frac{S\sqrt{\pi}}{2\beta\sqrt{A_c}} \quad [2.1]$$

Where S is the unloading stiffness (initial slope of the unloading load–displacement curve at the maximum displacement of penetration (or peak load)), A is the projected contact area between the tip and the substrate and β is a constant that depends on the geometry of the indenter (β = 1.167 for the Berkovich tip (Oliver and Pharr, 1992)). Conventional nanoindentation hardness refers to the mean contact pressure; this hardness, which is the contact hardness (H_c) is actually dependent upon the geometry of the indenter (Eq.2.2–2.4).

$$H_c = F / A \qquad [2.2]$$

where,

$$A(h_c) = 24,5h^2 + a_1 h + a_{1/2} h^{1/2} + \ldots + a_{1/16} h^{1/16} \qquad [2.3]$$

and

$$h_c = h_m - \varepsilon \frac{P_m}{S_m} \qquad [2.4]$$

where h_m is the total penetration displacement of the indenter at peak load, P_m is the peak load at the indenter displacement h_m, and ε is an indenter geometry constant, equal to 0.75 for the Berkovich indenter. Prior to indentation, the area function of the indenter tip was calibrated in a fused silica, a standard material for this purpose (Bei et al., 2005; Koumoulos et al., 2011; Charitidis, 2010b).

2.3 Results and discussion

2.3.1 *Input functions – load–displacement curves*

The relation (input function) of displacement change to time for the materials examined in this work is plotted in Fig. 2.2 (schematic trapezoidal load–time $P = P(t)$ input function). The loading–unloading curves of the probed materials are presented in Fig. 2.3 (comparison of all probed materials is presented, for an applied load of 5000 μN). Apart from Ni nanocomposite with TiO_2 filler particles, pure Ni exhibits higher resistance to applied load (higher applied load values

Figure 2.2 Schematic trapezoidal of load–time $P = P(t)$ function for nanoindentation experiment

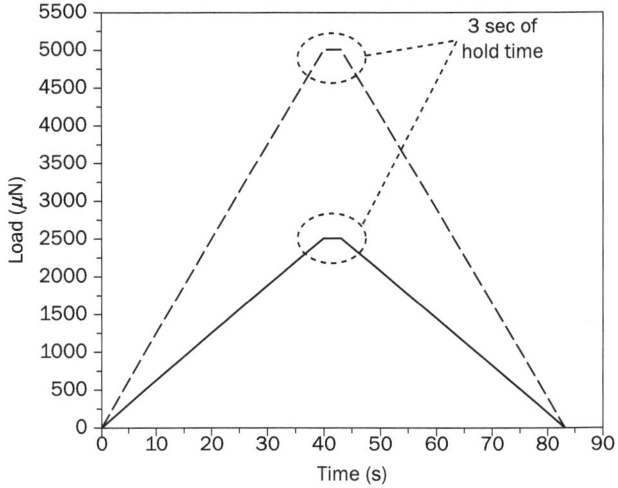

Figure 2.3 Loading–unloading curves of metals and alloys for applied load of 5000 μN

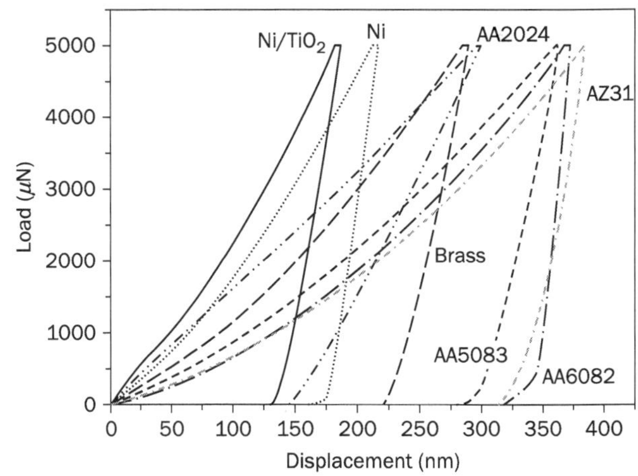

were needed for Ni to reach the same displacement as the rest of the materials). In the case of AA6082-T6, greater plasticity is revealed, i.e. energy stored in the material after the indentation is over (total integration of curve area), with

AA2024 exhibiting higher elastic recovery (Fig. 2.4) in the case of 5000 μN of applied load.

In Fig. 2.4, the comparison of plastic deformation for 2500 μN and 5000 μN reveals a range of difference of 5–10 per cent, implying enhancement of elasticity for lower

Figure 2.4 (a) Plasticity and (b) change of plasticity of metals and alloys for applied loads of 2500 and 5000 μN

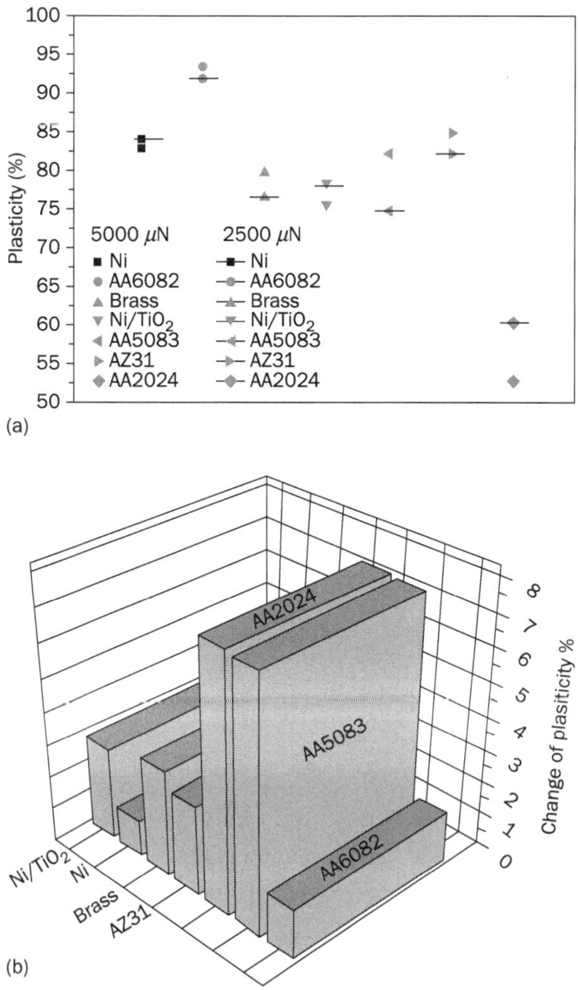

loads in the cases of AA6082, AA5083, brass and AZ31. This may be attributed to the real physical effect of superelastic behaviour of materials under microNewton-scale forces, due to the inactivation of dislocations; it may also be an artefact due to the piling-up of the surface during indentation. Pile-up is the tendency of softer materials to overflow plastically out of the indented region (further analysis below); this phenomenon is connected to the material, which is very soft, and also to the geometry of the tip (decreasing the corner angle of the tip increases the plastic deformation of the material and thus its pile-up) and it leads to an overestimation of *H* and *E* (underestimation of the contact area).

2.3.2 Pop-ins power law

Figure 2.5 shows that the loading-unloading curves for all samples exhibit interesting local discontinuities measured in

Figure 2.5 Discontinuities from indentations on metals and alloys in this study. Circles show 'pop-in' yield points.

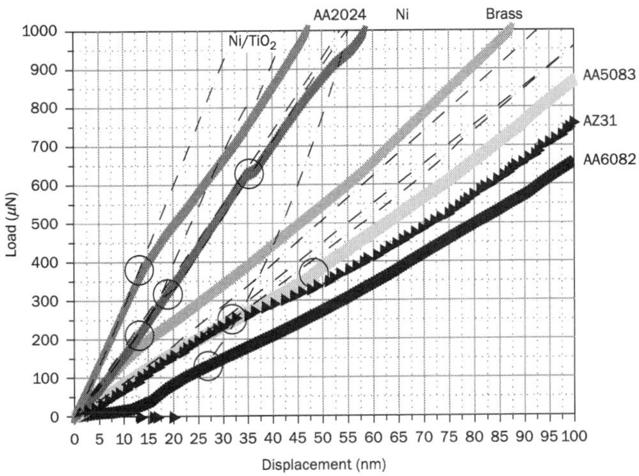

the load-controlled test of this work; these are characteristic of energy-absorbing or energy-releasing events occurring beneath the indenter tip. Three different physical phenomena usually occur in nanoindentation testing of metals of various states of bonding and structural order: dislocation activity during a shallow indentation, shear localization into 'shear bands', and phase transformation with a significant volume increase during unloading of indentation (Schuh, 2006).

Many materials undergo phase transformations when subjected to large hydrostatic stresses, and the pressure beneath a nanoindenter is generally quite high (on the order of several GPa) (Schuh, 2006). The association of pop-in events with the beginning of material phase transformation is simply the sudden extrusion of highly plastic transformed material from underneath the indenter. The sudden displacement discontinuities, i.e. the pop-ins, were observed in the loading part. The first pop-in (referred to as yield type pop-in in the literature) implies that strain is accommodated by an abrupt existence of atomic activity beneath the indenter, which could be attributed to activation of a dislocation source (Schuh, 2006).

Shallow displacements are equivalent to a linear elastic response; the curvature of the free energy diagram at an equilibrium position gives rise to a particular elastic constant, represented by the initial slope in a load displacement curve. Larger displacements elicit a non-linear response. However, one property of particular interest is the theoretical shear strength of the material: the limit where a further increase in displacement elicits no further increase in the restoring load. The general relationship between the applied load (P) and the penetration displacement (h) of an indenter may be described by power law as (Sneddon, 1948):

$$P = ah^b \qquad [2.5]$$

where the constants *a* and *b* are geometric and material parameters, respectively.

Load–displacement curves often reveal discontinuities or pop-ins in the loading part. Various examples are reported in the literature for metals, although a complete explanation of this behaviour is still being sought (Bahr et al., 1998; Kiely et al., 1998; Mann and Pethica, 1999). While yield points reported for Si and some other ceramics may be connected to phase transformations (when the mean contact pressure closely matches the critical pressure a possible structural transformation is triggered (Schuh, 2006; Gridneva et al., 1972; Gerk and Tabor, 1978; Johnson, 1970)), the exhibited yield points in most of the known metals clearly reveal the beginning of dislocation plasticity (the plastic deformation of metals occurs due to the motion of dislocations). Evaluation of the maximum stress under the tip reveals that stress values (almost equal to theoretical shear stress) occur in the surface of the metal (dislocation activity starts first at the yield type pop-in load) as the generation of dislocations in the nanoscale stressed volume forms the perfect crystal environment (Liang et al., 2003; Wo et al., 2005). As shown in Fig. 2.5, the transition from purely elastic to elastic/plastic deformation, i.e. gradual slope change (yield type pop-in) of all materials occurs in the load–displacement curves at approximately 10–30 nm.

The indenter displacement in most cases is accommodated plastically, and only a small portion is elastically recovered during unloading. Discontinuity in load–displacement curves, which is commonly referred to as a pop-in effect (Fig. 2.6), was observed frequently in aluminium, indicating a process of producing mobile dislocations. The initial pop-in is usually associated with homogeneous dislocation nucleation, while subsequent similar events often involve avalanches of dislocation activity (Liu et al., 2004).

Figure 2.6 **Pop-ins and elbows indicated in nanoindentation loading–unloading curves**

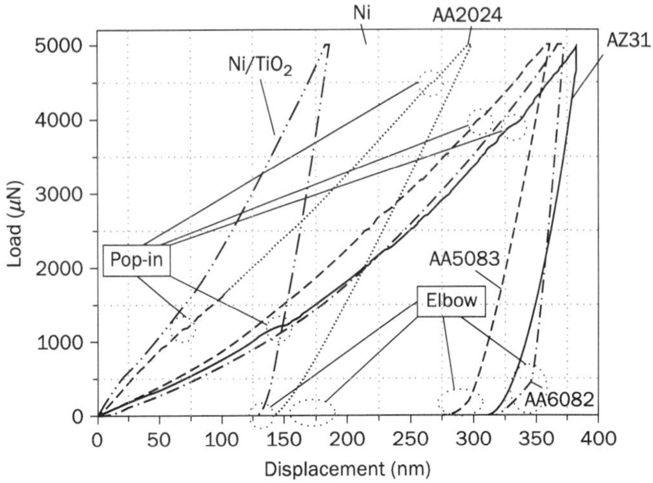

Gogotsi and co-workers (Domnich and Gogotsi, 2002; Gogotsi et al., 2000; Domnich et al., 2000) and Juliano et al. (2003) proposed that the 'pop-out' behaviour corresponds to the formation of metastable Si-XII/Si-III crystalline phases, in the case of silicon nanoindentation. High stresses can cause plastic deformation not only by dislocation activity, but also by pressure-induced phase transformations to denser crystalline and amorphous forms (Ge et al., 2004; Schmucker et al., 2003; Gilman, 1992; Domnich and Gogotsi, 2001). The transformation mechanisms are dependent on the indentation testing conditions, i.e. peak load and loading/unloading rate (Schmucker et al., 2003) or indenter angle (Jang et al., 2005). This results in a change in the unloading curve, either as a pop-out or 'elbow' phenomenon (Fig. 2.6), which indicates a lower contact depth, (h), and therefore may influence the calculation of H. Thus, the experimental measurement of the H of indented materials is slightly higher than that of the ideal value.

2.3.3 Pile-up/sink-in deformation

The contact area is influenced by the formation of pile-ups and sink-ins during the indentation process. To accurately measure the indentation contact area, pile-ups/sinks-ins should be appropriately accounted for. The presence of creep (time/rate dependent property of materials) during nanoindentation has an effect on pile-up, which results in incorrect measurement of the material properties. Fischer-Cripps (2004) observed this behaviour in aluminium, where the measured E was much less than expected (Fischer-Cripps, 2004). Rar et al. (2005) observed that the same material, when allowed to creep for a long duration, produced a higher value of pile-up/sink-in, indicating a switch from an initial elastic sink-in to a plastic pile-up (Rar et al., 2005). Significant pile-up forms for materials start from an h_c/h_m value of 0.7–0.88 (Khan et al., 2010).

In Fig. 2.7(a, b), the normalized pile-up/sink-in height h_c/h_m is plotted vs. the normalized hardness H/E^* for all samples (E^* is defined as reduced modulus (Oliver and Pharr, 1992)). It is reported that materials with high H/E^*, i.e. hard materials, undergo sink-in whereas materials pile-up with low H/E^*, i.e. soft materials, undergo pile-up. In general it is also observed that when H/E^* is high (hard materials), materials undergo sink-ins regardless of work hardening and strain rate sensitivity, and all materials collapse to a single curve. In addition, for softer materials with low H/E^*, pile-up depends on the degree of work hardening (Elmustafa, 2007). Softer materials, i.e. low H/E^*, possess a plastic zone, which is hemispherical in shape and meets the surface well outside the radius of the circle of contact and pile-up is expected. On the other hand, for materials with high values of H/E^* the plastic zone is contained within the boundary of

Figure 2.7 Normalised pile-up/sink-in height *h/hc* of metals and alloys vs. (a) displacement and (b) *H/E** ratio

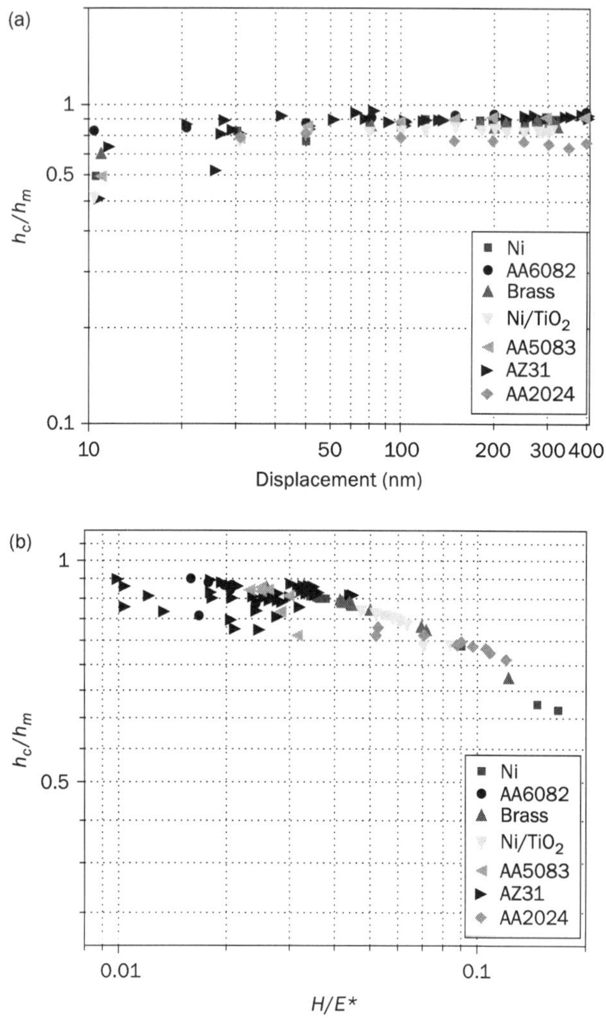

the circle of contact and the elastic deformations that accommodate the volume of indentation are spread at a greater distance from the indenter. In Fig. 2.8, pile-up of indented samples through SPM imaging is presented.

Figure 2.8 Pile-up of indented materials through SPM imaging for 2500 and 5000 μN

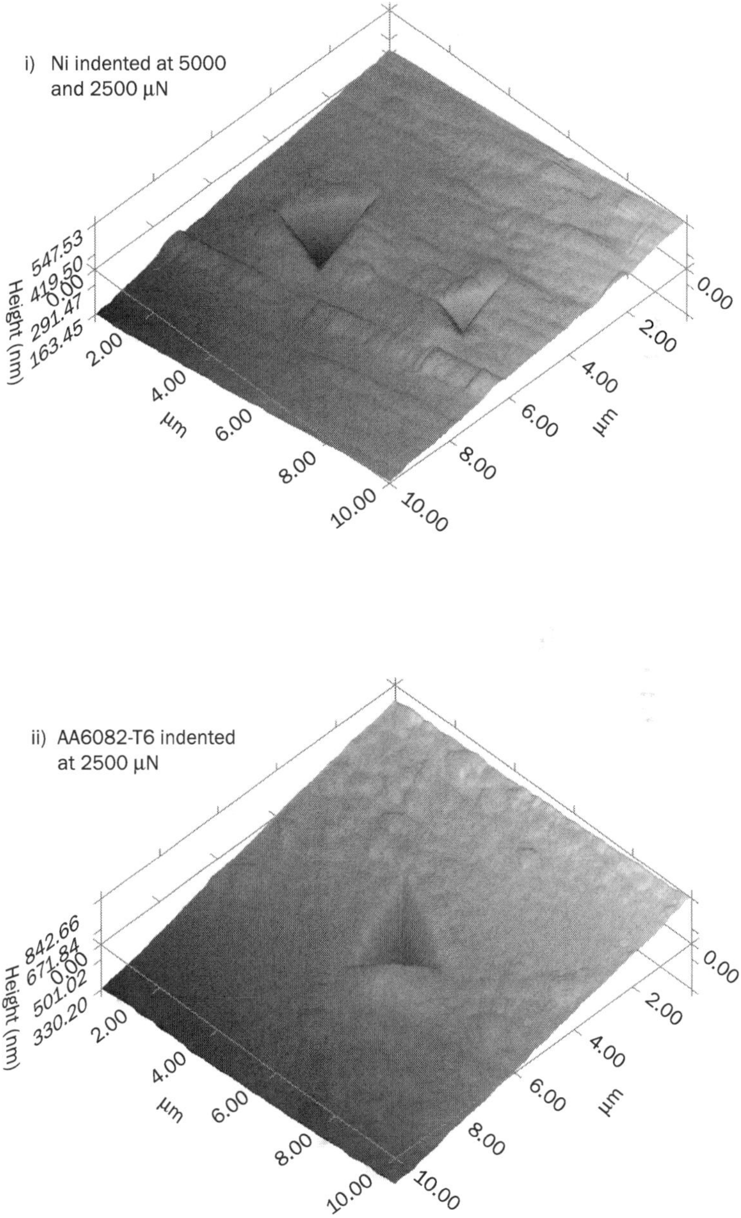

(*Continued*)

Figure 2.8 Pile-up of indented materials through SPM imaging for 2500 and 5000 μN (*Continued*)

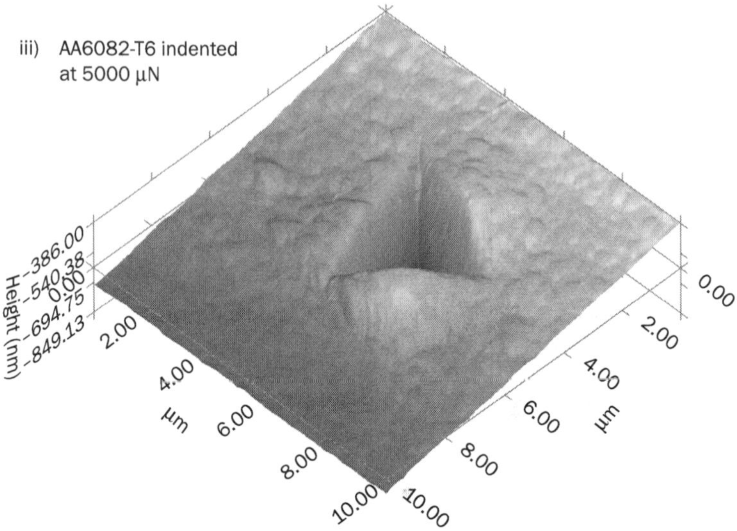

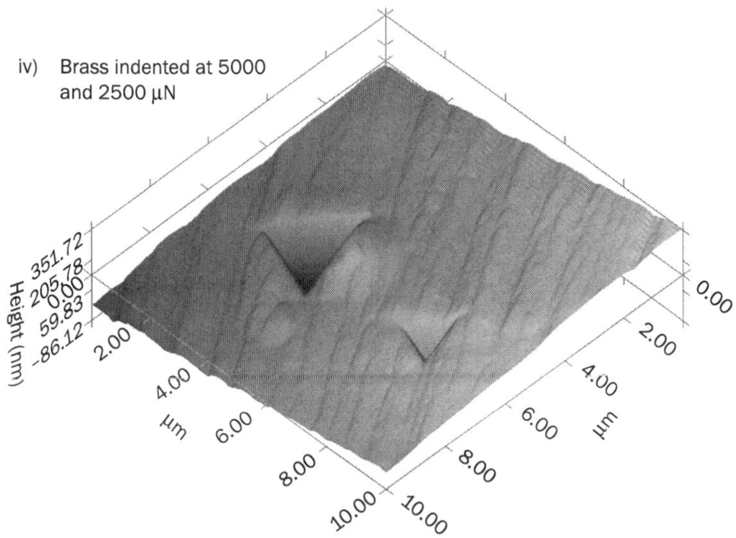

Figure 2.8 (*Continued*)

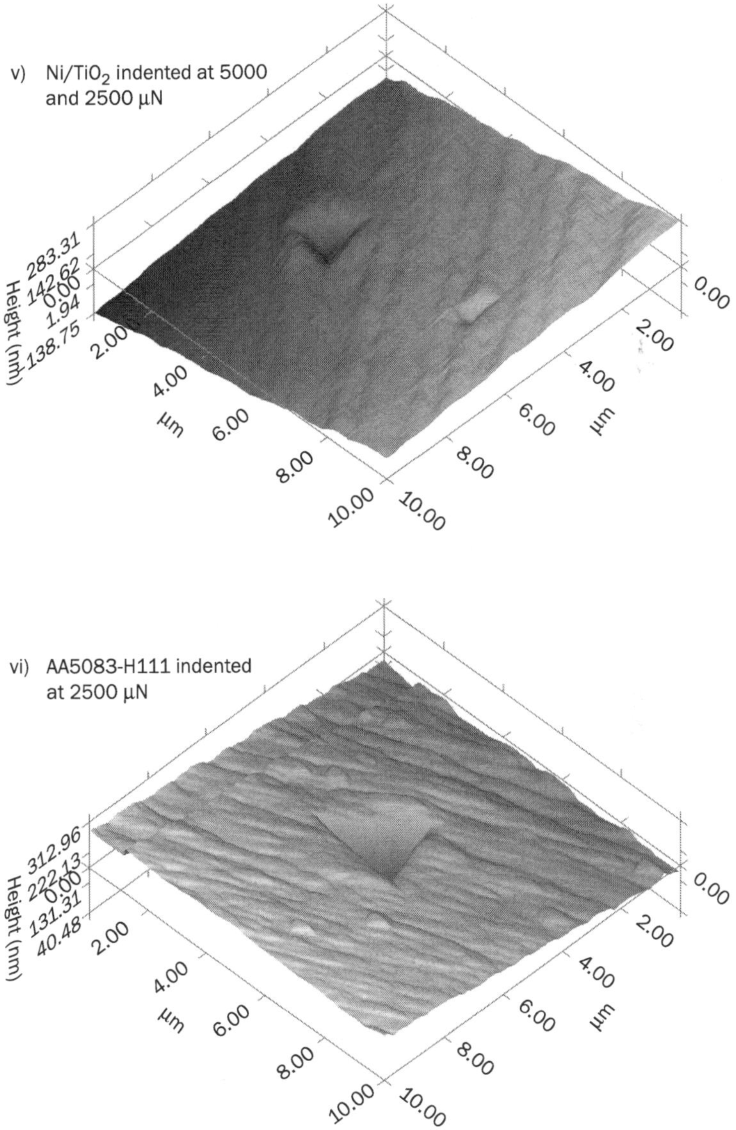

(*Continued*)

Figure 2.8 Pile-up of indented materials through SPM imaging for 2500 and 5000 μN (*Continued*)

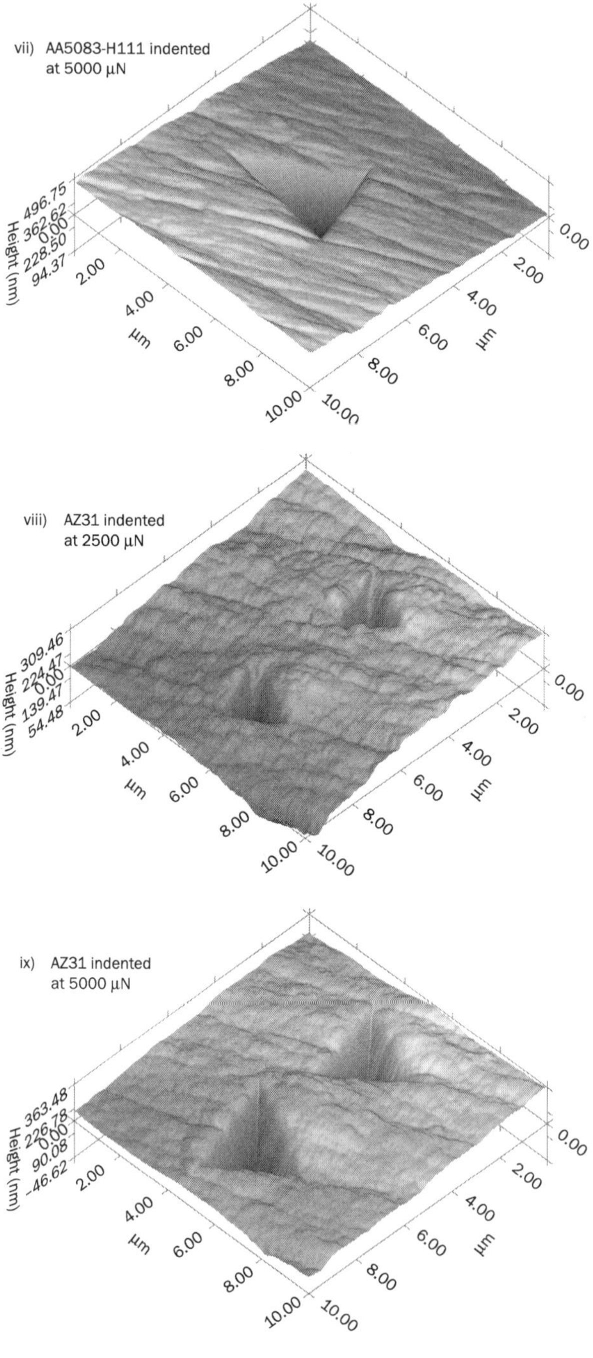

Figure 2.8 (*Continued*)

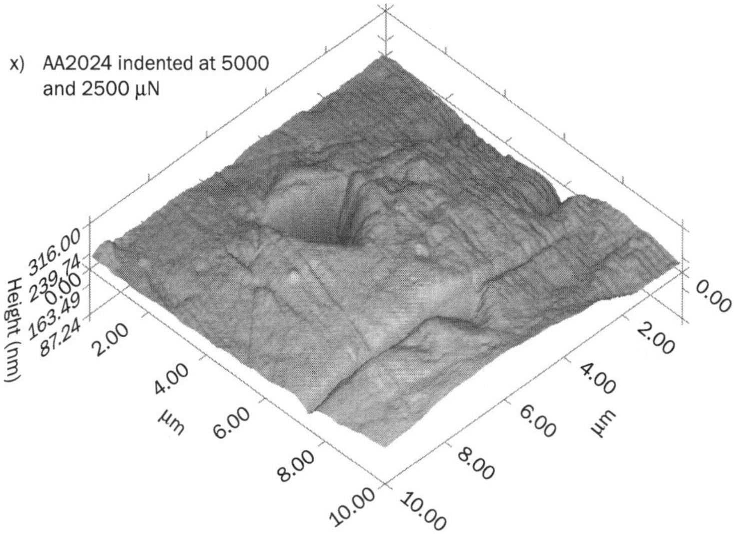

Higher stresses are expected in high H/E^* materials and high stress concentrations develop towards the indenter tip, whereas in the case of low H/E^* materials the stresses are lower and are distributed more evenly across the cross-section of the material (Gleiter, 2000). Rate sensitive materials experience less pile-up compared to rate insensitive materials due to strain hardening. Cheng and Cheng reported a 22 per cent pile-up for a work hardening exponent (Cheng and Cheng, 1998). This is consistent with the fact that when h_c/h_m approaches 1 for small H/E^*, deformation is intimately dominated by pile-up (Hill et al., 1989; Biwa and Storåkers, 1995). On the other hand, when h_c/h_m approaches 0 for large H/E^* it corresponds to purely elastic deformation and is apparently dominated by sink-in (pure Ni, Ni/TiO_2, AA2024) in a manner prescribed by Hertzian contact mechanics (Hertz, 1896).

The dependence on the elastic–plastic behaviour of the material is related to the response of the material being indented (McElhaney et al., 1998). The degree of sink-in or pile-up of the materials is reported to be expressed as a function of the work hardening exponent (Norbury and Samuel, 1928; McClintock and Rhee, 1962). During nanoindentation, materials with a low work hardening exponent (n) accommodate the volume of material, ejecting it to the sides of the tip (pile-up). In the same way, in materials with a high value of n ($n > 0.3$) the sink-in effect takes place. In both cases, the contact area is different from the cross-sectional area estimated by the method described by Oliver and Pharr (1992). Consequently, there is a deviation between the real and the computed area that is controlled by the elastic–plastic behaviour.

2.3.4 Stress–strain analysis

In the present study, nanoindentation data were also used for the extraction of stress–strain curves of the metals and alloys (Cao and Zhang, 2008). Through this analysis, plastic properties of materials can be defined. Cao and Zhang (2008), based on previous studies of Johnson (1970), Tabor (1951), Hill (1992), Bower et al. (1993), Storåkers and Larsson (1994) and others, presented the theoretical analysis of nanoindentation strain, taking into consideration that as the tip penetrates into the sample, its shape changes. At shallow indentation displacements, which are defined by the equation presented below (Eq. [2.6]), the indenter's shape is considered spherical, but as it further indents the sample the tip geometry is not negligible. The indentation stress is defined as the instant hardness or mean pressure underneath the indenter [Eq. 2.6]. Higher values of hardness are calculated at higher applied loads or when using sharper tips (Dao et al., 2001; Cao and Zhang, 2008).

The theoretical background of the indentation stress–strain analysis is based on the Oliver–Pharr method. The average stress, σ, which can be sustained under the indentation load (Freund and Suresh, 2004) is given by:

$$\sigma = \frac{P}{A} \qquad [2.6]$$

where P is the applied load and A is the area function of nanoindenter tip $A = f(h_c)$, where h_c is the contact depth, which is calculated through the Oliver–Pharr model. However, indentation strain determination is more complicated. When the indenter is considered spherical, the contact radius, a, is defined by:

$$a = \sqrt{\frac{A}{\pi}} = 0.5642\sqrt{A} \qquad [2.7]$$

and strain, ε, by:

$$\varepsilon_{Berkovich} = \frac{h_c}{a} \quad \text{for } h_c \leq \delta_o \qquad [2.8]$$

where δ_o is the maximum contact depth below which the indenter shape is spherical and is estimated by:

$$\delta_o = R(1 - \sin\theta) \qquad [2.9]$$

where θ is the semi-angle of the indenter tip and R is the curvature radius of the Berkovich indenter tip. As the indenter further penetrates the sample, the strain is calculated by:

$$\frac{d\varepsilon_{Berkovich}}{dt} = \frac{1}{h_c}\frac{dh_c}{dt} \quad \text{for } h_c \geq \delta_o \qquad [2.10]$$

A full nanoindentation stress–strain curve can be obtained by combining Eq. [2.6] and [2.8] for $h_c \leq \delta_o$ and Eq. [2.10] for $h_c \geq \delta_o$ (Fig. 2.9). Ni containing TiO_2 exhibits higher stress than expected, while pure Ni yields at an earlier stage

Figure 2.9 (a) Stress–strain curves and (b) yielding of stress–strain curves obtained from nanoindentation data for 5000 μN of applied load

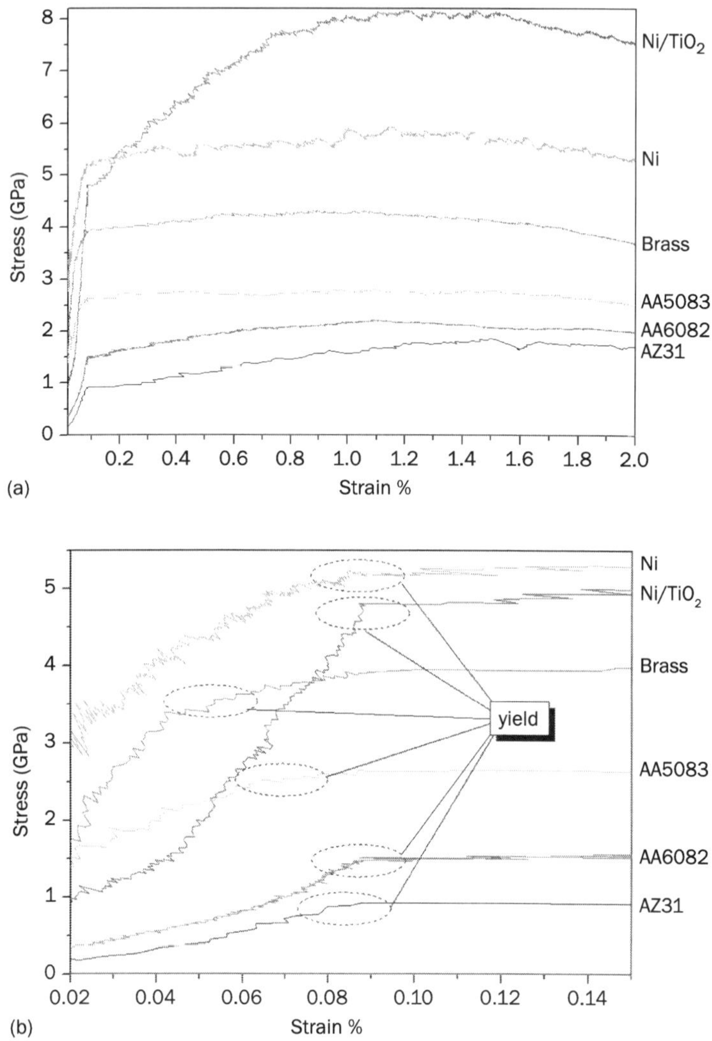

(while brass exhibits higher stress than aluminium and magnesium alloys).

The tensile curves of the samples are plotted in Fig. 2.10, according to literature studies (a broad range of strengths is

Figure 2.10 (a) Stress–strain curves and (b) yielding of stress–strain curves obtained from the literature for metals and alloys

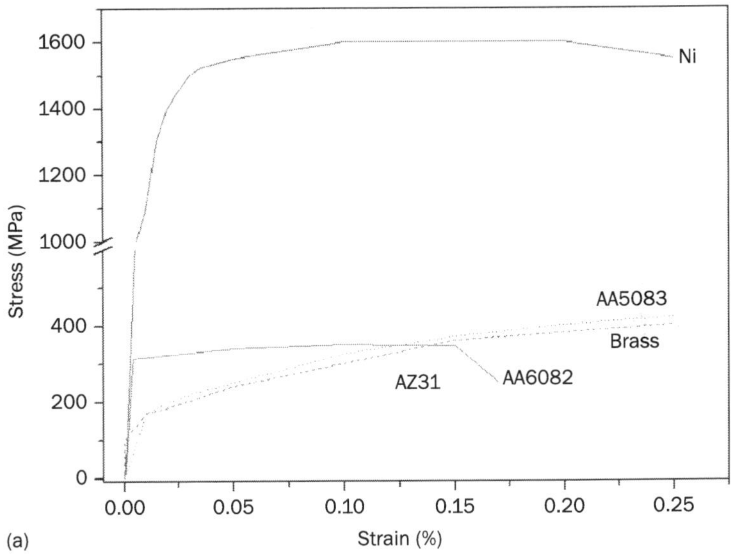

(a)

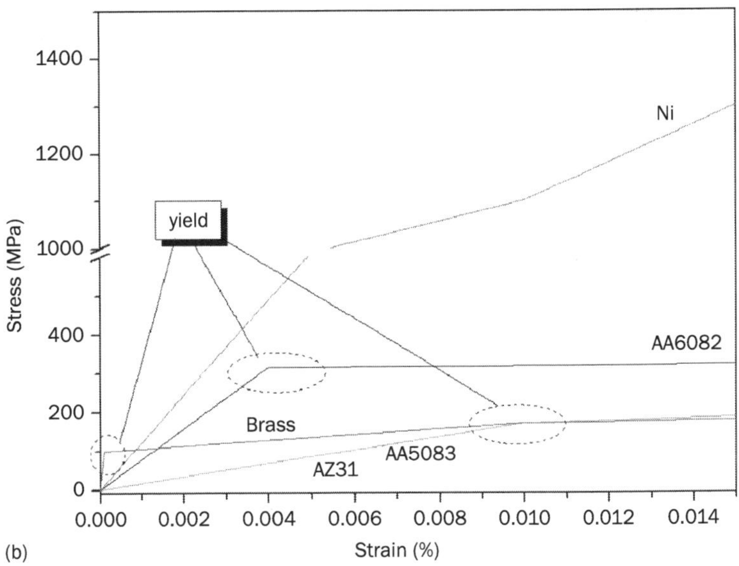

(b)

obtained). The softest materials are the AA5083-H111 Al-alloy and the brass material (Ni is the hardest as expected, due to fine granulometry structure).

It is important to note that nanoindentation and tensile tests do not measure the same properties. Generally, nanoindentation modulus (elastic modulus) values using the standard Oliver–Pharr method are higher than corresponding results from standard tensile tests. Several factors have been suggested as possible explanations for this: Surface effects (the initial part of the load–displacement curve is affected by surface roughness, surface oxidation (Koumoulos et al., 2011) and other surface phenomena), the type of loading is compressive in nanoindentation (not tensile), the test frequencies are quite different (70 Hz for nanoindentation, compared to a much lower frequency in the tensile test) and hydrostatic pressure generated below the Berkovich indenter tip (Charitidis, 2010b).

2.3.5 Depth profiling of metals and alloys

At each imposed displacement, the true indentation *H* and *E*, can be deduced from the curves (Oliver and Pharr, 1992). The graphs in Figs. 2.11 and 2.12 show the mean value of the *H* measurements as a function of the imposed displacement (the trend of the curvatures is similar for all samples). As the indentation displacement decreases below 100 nm, a rapid increase of the *H* value is observed. This rapid increase is probably a combination of either the real effect of a native oxide at the surface or an effect of the polishing procedure, and an artefact of the shape of the indenter tip for shallow displacements (Koumoulos et al., 2011; Charitidis, 2010b; Liu and Ngan, 2001; Lucas and Chevallier, 1994).

The nanohardness data in Fig. 2.11 and 2.12 were calculated using the Oliver–Pharr method which does not

Figure 2.11 Hardness of Ni, brass and Ni/TiO_2 for various displacements

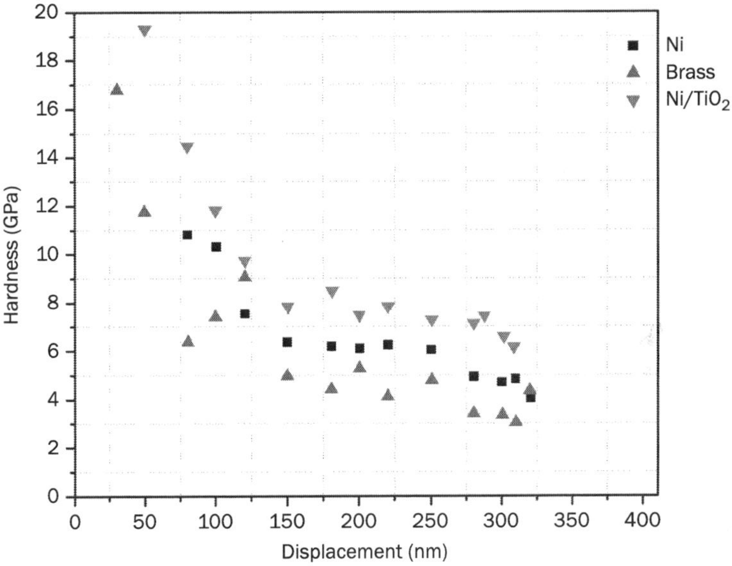

Figure 2.12 Hardness of AA6082-T6, AA5083-H111, AZ31 and AA2024 for various displacements

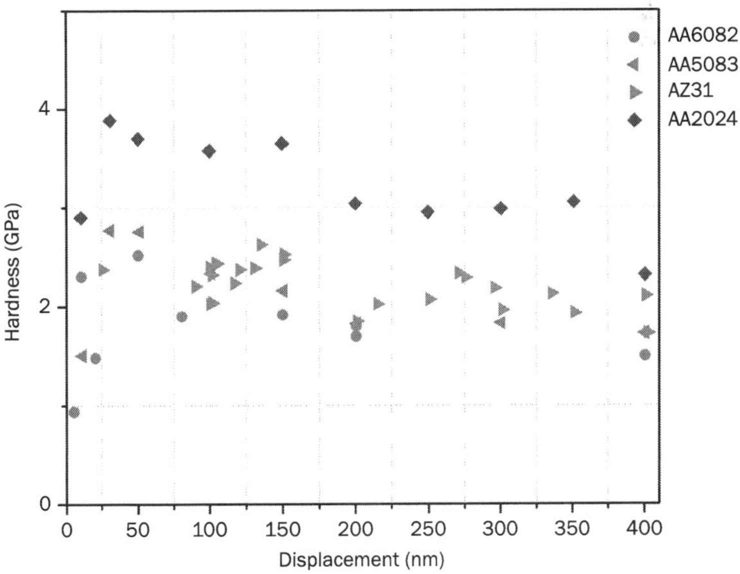

take into account elastic–plastic effects. Göken et al. (2001) proposed a method of correcting pile-up effects and possible surface roughness. This method allows determination of a correction factor for *H* based on the relation between the indentation modulus and Young's modulus (Vlassak and Nix (1994) reported that *E* is highly dependent on the crystallographic orientation but *H* is not).

In Figs. 2.13 and 2.14, the mean value of the *E* is plotted versus the displacement of the indentation. For aluminium alloys, an almost constant value of *E* is obtained over the different applied test conditions, which seems reasonable. A range of values of ~50–100 GPa is obtained, which is in fairly good agreement with the reported Young's modulus in the literature (Callister, 1990). The same fact is confirmed for brass which leads to a value of ~100 GPa, relatively close to the 110 GPa reported (Gil Sevillano et al., 2000).

Figure 2.13 Elastic modulus of Ni, brass and Ni/TiO_2 for various displacements

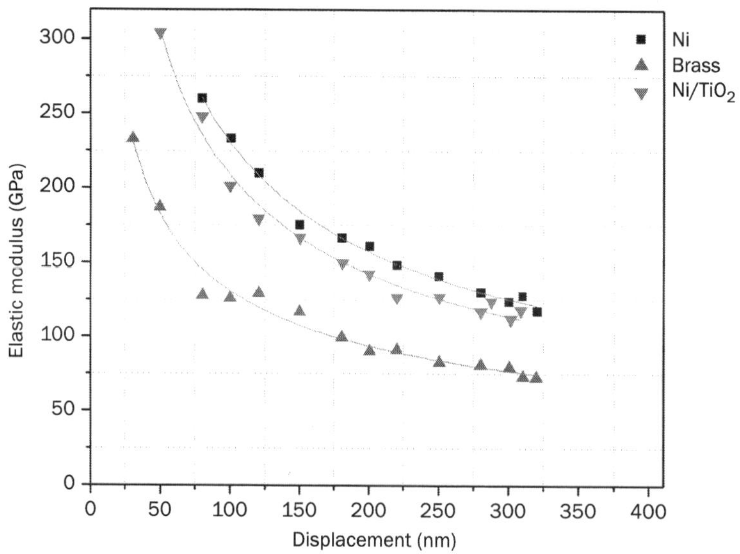

Figure 2.14 Elastic modulus of AA6082-T6, AA5083-H111, AZ31 and AA2024 for various displacements

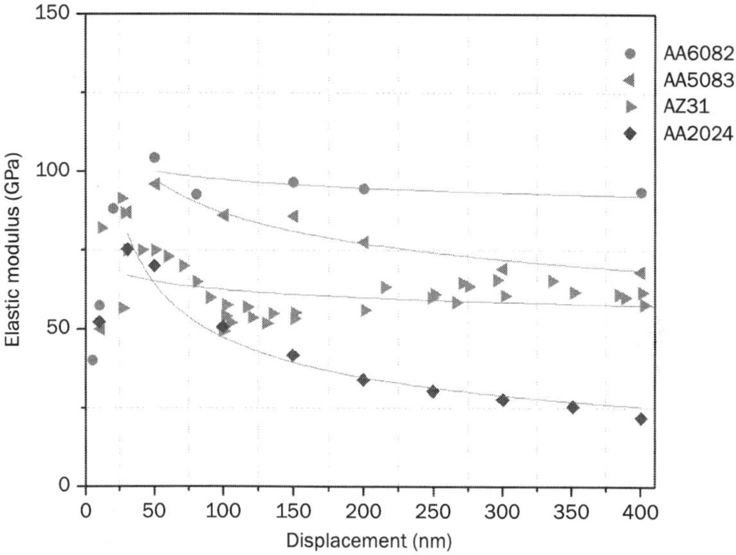

The deviation from the standard values is attributed to the *E* values, which are calculated by using the contact area from the indentation displacement considered since the first contact point with the initial flat surface, as described by the Oliver–Pharr method. However, the plastically deformed zone around the indented area can pile-up or sink-in against the indenter, depending on the material's work hardening. The consequence of this behaviour is a slight deviation in the standard values of the elastic modulus (Rodriguez and Gutierrez, 2003). As far as brass is concerned, Vlassak and Nix (1994) had determined 25 per cent of scatter between indentation modulus for different crystallographic orientations, but even higher discrepancies can be found, depending on the material.

In Fig. 2.15, the square of the *H* value obtained in the indentation tests is plotted as a function of the reciprocal of

Figure 2.15 (a) Square of the nanohardness value against the inverse of the displacement; (b) comparison of hardness at 400 nm and extrapolated hardness; and (c), (d) hardness change of metals and alloys

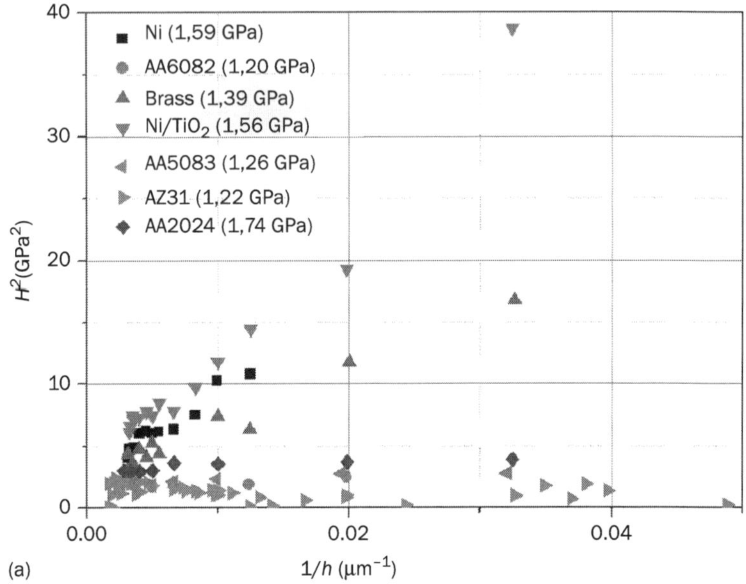

(a)

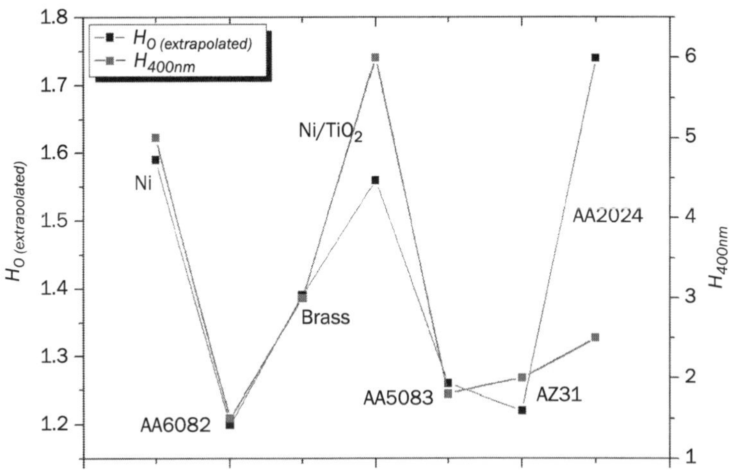

(b)

Figure 2.15 (*Continued*)

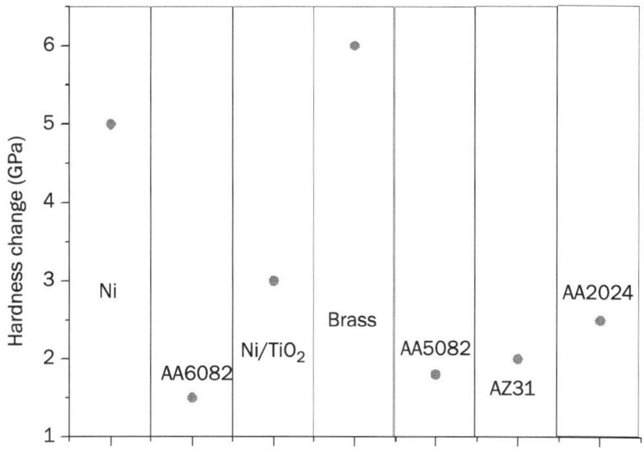

(c)

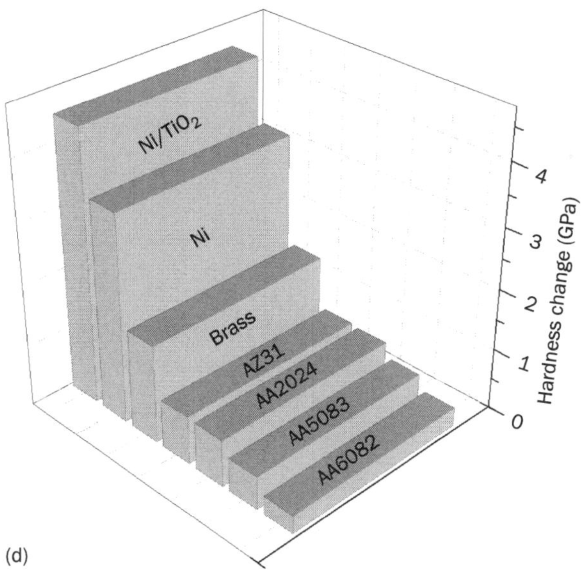

(d)

the indentation displacement. It can be seen that a linear relationship is closely followed for the majority of the materials, in agreement with the literature (Rodriguez and Gutierrez, 2003). The *H* value at infinite displacement,

H_0, can be estimated by extrapolating the aforementioned linear relationships to $1/h = 0$. The obtained values are shown in the data embedded in Fig. 2.15(a). The comparison of H_0 and H at 400 nm is presented in Fig. 2.15(b), where H at infinite displacement clearly matches with H at 400 nm for AA6082-T6, brass and AA5083-H111. However, the same comparison for pure Ni, Ni with TiO_2 and AA2024 exhibits great deviation (reduced plastic deformation in higher applied loads, dominated by sink-in), revealing that in order to reach constant nanomechanical properties (of bulk material), indenting to greater displacement is needed.

2.4 Conclusions

Pure Ni exhibited higher resistance to applied load (higher applied load values were needed for Ni to reach the same displacement as the rest of the materials). In the case of AA6082-T6, greater plasticity was revealed, i.e. energy stored in the material after the indentation was complete (total integration of curve area), with AA2024 exhibiting higher elastic recovery in the case of 5000 μN of applied load. The comparison of plastic deformation at 2500 μN and 5000 μN loading revealed a difference of between 5 and 10 per cent, implying an enhancement of elasticity for lower loads in the cases of AA6082, AA5083, brass and AZ31. All examined materials exhibited interesting local discontinuities measured in the load-controlled test, which are characteristic of energy-absorbing or energy-releasing events occurring beneath the indenter tip. In the load–displacement curves, brass exhibited earlier transition from purely elastic to elastic/plastic deformation i.e. gradual slope change (yield type pop-in), followed by pure Ni and Ni with TiO_2 (AZ31 and aluminium alloys exhibited late transition). Aluminium alloys (AA5083 and AA6082) exhibited significant elbow effect in the unloading phase. Hardness at ~400 nm and extrapolated hardness showed almost the same behaviour for examined metals and alloys (in agreement with similar studies in the literature). Comparison of both *H* values of pure Ni, Ni with TiO_2 and AA2024 exhibited great deviation (reduced plastic deformation at higher applied loads, dominated by sink-in), revealing that in order to reach constant nanomechanical properties (of bulk material), indenting to greater displacement is needed.

2.5 References

Atkinson M (1991), 'Origin of the size effect in indentation of metals', *Int J Mech Sci*, **33**: 843–50.

Bahr D F, Kramer D E and Gerberich W W (1998), 'Non-linear deformation mechanisms during nanoindentation', *Acta Mater*, **46**: 3605–17.

Bei H, George E P, Hay J L and Pharr G M (2005), 'Influence of indenter tip geometry on elastic deformation during nanoindentation', *Phys Rev Lett*, **95**: 045501 (4 pages).

Biwa S and Storåkers B (1995), 'Analysis of fully plastic Brinell indentation', *J Mech Phys Solids*, **43**: 1303–33.

Bower A F, Fleck N A, Needleman A and Ogbonna N (1993), 'Indentation of a power law creeping solid', *P R Soc London A Mat*, **441**: 97–123.

Bucaille J L, Stauss S, Felder E and Michler J (2003), 'Determination of plastic properties of metals by instrumented indentation using different sharp indenters', *Acta Mater*, **51**: 1663–78.

Callister W D (1990), *Materials Science and Engineering*. New York: Wiley.

Cao Y P and Lu J (2005), 'Size-dependent sharp indentation-II: a reverse algorithm to identify plastic properties of metallic materials', *J Mech Phys Solids*, **53**: 49–62.

Cao Z and Zhang X (2008), 'Nanoindentation stress–strain curves for plasma-enhanced chemical vapour deposited silicon oxide thin films', *Thin Solid Films*, **516**: 1941–51.

Cavaliere P (2009), 'Fatigue properties and crack behavior of ultra-fine and nanocrystalline pure metals', *Int J Fatigue*, **31**: 1476–89.

Charitidis C A (2010), 'Nanomechanical and nanotribological properties of carbon-based thin films: A review', *Int J Refract Met H*, **28**: 51–70.

Charitidis C A (2010), 'Local mechanical properties and nanoscale deformation of aluminum and magnesium lightweight alloys', *Int J Surf Sci Eng*, in press.

Cheng Y T and Cheng C M (2004), 'Scaling, dimensional analysis, and indentation measurements', *Mat Sci Eng R*, **44**: 91–149.

Cheng Y T and Cheng C M (1998), 'Effects of "sinking in" and "piling up" on estimating the contact area under load in indentation', *Phil Mag Lett,* **78**: 115–20.

Dao M, Chollacoop N, van Vliet K J, Venkatesh T A and Suresh S (2001), 'Computational modeling of the forward and reverse problems in instrumented sharp indentation', *Acta Mater*, **49**: 3899–918.

Doerner M F and Nix W D (1998), 'Stresses and deformation processes in thin films on substratcs', *Crit Rev Solid State*, **14**: 225–68.

Domnich V and Gogotsi Y (2002), 'Phase transformations in silicon under contact loading', *Rev Adv Mater Sci*, **3**: 1–36.

Domnich V and Gogotsi Y (2001), *Handbook of Surface and Interface of Materials*. New York: Academic Press.

Domnich V, Gogotsi Y and Dub S (2000), 'Effect of phase transformations on the shape of the unloading curve in the nanoindentation of silicon', *Appl Phys Lett*, **76**: 2214–6.

Elmustafa A A (2007), 'Pile-up/sink-in of rate-sensitive nanoindentation creeping solid', *Model Simul Mater Sc*, **15**: 823–34.

Elmustafa A A and Stone D S (2003), 'Nanoindentation and the indentation size effect: kinetics of deformation and strain gradient plasticity', *J Mech Phys Solids*, **51**: 357–81.

Fischer-Cripps A C (2004), 'A simple phenomenological approach to nanoindentation creep', *Mater Sci Eng A-Struct*, **385**: 74–82.

Freund L B and Suresh S (2004), *Thin Film Materials: Stress, Defect Formation and Surface Evolution.* New York: Cambridge University Press.

Ge D, Domnich V, Juliano T, Stach E A and Gogotsi Y (2004), 'Structural damage in boron carbide under contact loading', *Acta Mater*, **52**: 3921–7.

Gerk A P and Tabor D (1978), 'Indentation hardness and semiconductor-metal transformation of germanium and silicon', *Nature*, **271**: 732–3.

Gilman J J (1992), 'Insulator–metal transitions at microindentations', *J Mater Res*, **7**: 535–8.

Gil Sevillano J, Buessler P, Vrieze J, Kaluza W, Bouaziz O et al. (2000), ECSC Steel RTD Final report, CECA7210-PR-044.

Gleiter H (2000), 'Nanostructured materials: basic concepts and microstructure', *Acta Mater*, **48**: 1–29.

Gogotsi Y G, Domnich V, Dub S N, Kailer A and Nickel K G (2000), 'Cyclic nanoindentation and Raman microspectroscopy study of phase transformations in semiconductors', *J Mater Res*, **15**: 871–9.

Göken M, Kempf M and Nix W D (2001), 'Hardness and modulus of the lamellar microstructure in PST-Ti-Al studied by nanoindentations and AFM', *Acta Mater*, **49**: 903–11.

Gridneva I V, Milman Y V and Trefilov M (1972), 'Phase transition in diamond-structure crystals during hardness measurements', *Phys Status Solidi A*, **14**: 177–82.

Hayes R W, Witkin D, Zhou F and Lavernia E J (2004), 'Deformation and activation volumes of cryomilled ultrafine-grained aluminum', *Acta Mater*, **52**: 4259–71.

Hertz H (1896), *Miscellaneous Papers*. London: Macmillan.

Hill R (1992), 'Similarity analysis of creep indentation tests', *P R Soc London A Mat*, **436**: 617–30.

Hill R, Storåkers B and Zdunek A B (1989), 'A theoretical study of the Brinell hardness test', *P R Soc London A Mat*, **423**: 301–30.

Hutchinson J W (2000), 'Plasticity at the micron scale', *Int J Solids Struct*, **37**: 225–38.

Jang J I, Lance M J, Wen S Q, Tsui T Y and Pharr G M (2005), 'Indentation-induced phase transformations in silicon: influences of load, rate and indenter angle on the transformation behaviour', *Acta Mater*, **53**: 1759–70.

Jeong D H, Gonzalez F, Palombo G, Aust K T and Erb U (2001), 'The effect of grain size on the wear properties of electrodeposited nanocrystalline nickel coatings', *Scripta Mater*, **44**: 493–9.

Jia D, Ramesh K T and Ma E (2003), 'Effects of nanocrystalline and ultrafine grain sizes on constitutive behavior and shear bands in iron', *Acta Mater*, **51**: 3495–509.

Johnson K L (1970), 'The correlation of indentation experiments', *J Mech Phys Solids*, **18**: 115–26.

Juliano T, Gogotsi Y and Domnich V (2003), 'Effect of indentation unloading conditions on phase transformation induced events in silicon', *J Mater Res*, **18**: 1192–201.

Khan M K, Hainsworth S V, Fitzpatrick M E and Edwards L (2010), 'A combined experimental and finite element approach for determining mechanical properties of aluminium alloys by nanoindentation', *Comp Mater Sci*, **49**: 751–60.

Kiely J D, Hwang R Q and Houston J E (1998), 'Effect of surface steps on the plastic threshold in nanoindentation', *Phys Rev Lett*, **81**: 4424–7.

King R B (1987), 'Elastic analysis of some punch problems for a layered medium', *Int J Solids Struct*, **23**: 1657–64.

Koumoulos E P, Charitidis C A, Daniolos N M and Pantelis D I (2011), 'Nanomechanical properties of friction stir welded AA6082-T6 aluminum alloy', *Mater Sci Eng B*, Article in Press.

Kumar K S, Van Swygenhoven H and Suresh S (2003), 'Mechanical behavior of nanocrystalline metals and alloys', *Acta Mater*, **51**: 5743–74.

Li H, Chosh A, Han Y H and Bradt R C (1993), 'The frictional component of the indentation size effect in low hardness testing', *J Mater Res*, **8**: 1028–32.

Liang H Y, Woo C H, Huang H, Ngan A H W and Yu T X (2003), 'Dislocation nucleation in the initial stage during nanoindentation', *Philos Mag*, **83**: 3609–22.

Liu R P, Herlach D M, Vandyoussefi M and Greer A L (2004), 'Undercooling and solidification of Al-50 at % Si alloy by electromagnetic levitation', *Metall Mater Trans A*, **35**: 607–12.

Liu Y and Ngan A H W (2001), 'Depth dependence of hardness in copper single crystals measured by nanoindentation', *Scripta Mater*, **44**: 237–41.

Lucas S and Chevallier J (1994), 'Nanohardness and transmission electron microscopy study of nitrogen-implanted aluminum', *Surf Coat Tech*, **65**: 128–32.

Mann A B and Pethica J B (1999), 'The effect of tip momentum on the contact stiffness and yielding during nanoindentation testing', *Philos Mag A*, **79**: 577–92.

Masumura R A, Hazzledine P M and Pande C S (1998), 'Yield stress of fine grained materials', *Acta Mater*, **46**: 4527–34.

McClintock F A and Rhee S S (1962), in: *Proceedings of the 4th US National Conference, Applied Mechanics 2*, p. 1007. Berkeley: ASME.

McElhaney K W, Vlassak J J and Nix W D (1998), 'Determination of indenter tip geometry and indentation contact area for depth-sensing indentation experiments', *J Mater Res*, **13**: 1300–6.

Miyahara K, Matsuoka S and Hayshi T (2001), 'Nanoindentation as a strength probe – a study on the hardness dependence of indent size for fine-grained and coarse-grained ferritic steel', *Metall Mater Trans A*, **32**: 761–768.

Nix W D and Gao H (1998), 'Indentation size effects in crystalline materials: A law for strain gradient plasticity', *J Mech Phys Solids*, **46**: 411–25.

Norbury A L and Samuel T (1928), 'The recovery and sinking-in or piling-up of material in the Brinell test, and the effects of these factors on the correlation of the Brinell with certain other hardness tests', *J Iron Steel I*, **117**: 673–87.

Oliver W C and Pharr G M (1992), 'An improved technique for determining hardness and elastic modulus using load and displacement sensing indentation experiments', *J Mater Res*, **7**: 1564–83.

Pelletier H, Krier J, Cornet A and Millc P (2000), 'Limits of using bilinear stress-strain curve for finite element modeling of nanoindentation response on bulk materials', *Thin Solid Films*, **379**: 147–55.

Rar A, Sohn S, Oliver W C, Goldsby D L, Tullis T E and Pharr G M (2005), *On the Measurement of Creep by Nanoindentation with Continuous Stiffness Techniques*. Boston: Materials Research Society.

Rodriguez R and Gutierrez I (2003), 'Correlation between nanoindentation and tensile properties: Influence of the indentation size effect', *Mat Sci Eng A-Struct*, **361**: 377–84.

Schmucker M, Schneider H and Kriven W M (2003), 'Indentation-induced amorphization in mullite single crystals', *J Am Ceram Soc*, **86**: 1821–2.

Schuh C A (2006), 'Nanoindentation studies of materials', *Mater Today*, **9**: 32–9.

Schwaiger R, Moser B, Dao M, Chollacoop N and Suresh S (2003), 'Some critical experiments on the strain-rate sensitivity of nanocrystalline nickel', *Acta Mater*, **51**: 5159–72.

Sneddon I N (1948), 'Boussinesq's problem for a rigid cone', *Math Proc Cambridge*, **44**: 492–507.

Storåkers B and Larsson P L (1994), 'On Brinell and Boussinesq indentation of creeping solids', *J Mech Phys Solids*, **42**: 307–32.

Suryanarayana C (2005), 'Recent developments in nanostructured materials', *Adv Eng Mater*, **7**: 983–92.

Tabor D (1951), *The Hardness of Metals*. Oxford: Clarendon Press.

Tymiak N I, Kramer D E, Bahr D F, Wyrobek T J and Gerberich W W (2001), 'Plastic strain and strain gradients at very small indentation depths', *Acta Mater*, **49**: 1021–34.

Vlassak J J and Nix W D (1994), 'Measuring the elastic properties of anisotropic materials by means of indentation experiments', *J Mech Phys Solids*, **42**: 1223–45.

Wang Y S, Qu S L, Gai Y X, Dong S and Liang Y C (2009), 'Residual strains of aluminum alloy characterized by nanoindentation', *T Nonferr Metal Soc*, **19**: s767–71.

Wo P C, Zuo L and Ngan A H W (2005), 'Time-dependent incipient plasticity in Ni_3Al as observed in nanoindentation', *J Mater Res*, **20**: 489–95.

3

Analysis of large-strain microindentation of crystalline polymers

Y. Lu, University of Kentucky, USA

Abstract: This chapter presents the analysis of large-strain microindentation tests for measuring the plastic properties of crystalline polymers in small volumes. Experimental results are obtained by using cylindrical, flat-ended indenters with the diameters of less than 100 μm (larger than the molecular dimensions of the crystalline polymers). The deformation process is analyzed theoretically to obtain material properties from the microindentation tests.

Key words: large-strain indentation test, plastic deformation, crystalline polymers.

3.1 Introduction

Indentation testing has been used for many years to measure the plastic properties of materials, and has been known as 'hardness testing'. The standard technique involves pushing a hard tip, typically a sphere (Brinell indenter), a cone (Rockwell indenter) or a diamond pyramid (Vickers indenter), into the material with a prescribed fixed load, and measuring either the depth or the diameter of the residual impression, either the depth or diameter (O'Neill, 1967). Such tests on polymers have been reported in the literature. However, the interpretation of the results was inexact because the displacement rate or loading rate was not controlled. Only qualitative information could be obtained about the material properties.

A more sophisticated form of indentation test, namely microindentation or nanoindentation, has been developed during the last two decades and offers significant advantages over a traditional hardness test. In this test, an indenter tip is pushed into the surface of the material under precise load or displacement control. The extremely small forces and displacements are recorded continuously throughout the loading–unloading cycle. The load–displacement curve can be analyzed to obtain material properties which extend beyond that obtainable from the standard hardness test (Loubet et al., 1984; Marshall, 1984; Doerner and Nix, 1986; Oliver and Pharr, 1992; Mayo and Nix, 1988; Ogilvy, 1993; Jervis and Spaepen, 1994; Pharr, Harding and Oliver, 1994; Vlassak and Nix, 1993 and 1994; Gubicza et al., 1996; Zhang and Mahdi, 1996; Fischer-Cripps, 1997; Jayaraman et al., 1998; Lu and Shinozaki, 2008; Lu et al., 2010).

The renewed interest in small-scale indentation has been driven by two factors: the improvements in instrumentation and the modern interest in microscopic structures. The development of the scanning tunneling microscope, and the associated atomic force microscope, has spurred interest in measuring extremely small forces and displacements, on the scale related to small groups of atoms. The improvements in piezoelectric positioning instrumentation that allows for atomically precise displacement control, and the need to measure small contact forces found with atomic force microscope tips, have made the development of micro- and nanoindentation instruments possible. A desire to probe the mechanical properties and understanding the failures of small-scale structures has spurred interest in understanding micro- and nanoindentation testing.

So far, most of the work has been related to the measurements of elastic properties of hard materials with sharp indenters. The deformation of purely elastic indentation is generally simple and the stress fields around the indenters of various geometry have been formulated in closed forms. The micro- and nanoindentation of elastic–plastic solids has gained considerable attention recently, with the purpose of determining the plastic characteristics of the materials such as yield strength, work hardening rate, residual stress, and even stress–strain curves (Swain, 1998; Swadener et al., 2001; Anton et al., 1999; Park and Pharr, 2004; Herbert et al., 2001; Lu et al., 2008). Most of the work involves the uses of indenters of parabolic shapes, i.e., spherical and conical indenters. The main advantage of using parabolic shaped indenters (compared to a sharp indenter such as pyramidal indenter) is that the indentation strain increases with the depth of penetration so that the deformation changes distinctly from purely elastic in the initial contact to fully plastic in the later stage (Park and Phar, 2004; Lu et al., 2008). The shape of the plastic zone under spherical or conical indenters has been assumed to be hemispherical. Johnson thus proposed that the indentation of an elastic–plastic solid is equivalent to a spherical or cylindrical cavity expanding in an infinite medium (Johnson, 1994). From the model, the mean indentation pressure (hardness) P_m can be predicted, and it has been related to the tensile yield strength σ_y by a constrain factor.

The present work analyzes the deformation in semi-crystalline polymers near the penetration with a flat cylindrical punch. A microindenter tip (less than 100 μm in diameter) is used in the experiments. The size of the tip is much larger than the molecular dimensions (the sizes of the spherulites) of the semi-crystalline polymers. The work extends earlier small displacement indentation studies in which the deformation field was mainly elastic, to large displacements with a typical normalized depth to one indenter diameter. At these large displacements, the material near the indenter tip is plastically deformed to large strains.

3.2 Experimental data

The large displacement microindentation experiment can be schematically described as shown in Figure 3.1. The test is run typically to an indentation depth (*d*) greater than at least one indenter diameter (2*a*), where *a* is the indenter radius. Figure 3.2 shows the deformed volume near indentation for a high density polyethylene (HDPE) by using a cylindrical flat-ended tip. The HDPE is a semi-crystalline polymer with a fine spherulite size of 5 μm in diameter. The crystallinity of the HDPE is 76 per cent as measured by differential scanning calorimetry (DSC). The image was obtained through transmitted light optical microscopy of a thin HDPE section cut from the central plane across the cylindrical indentation using a rotary microtome. Under crossed polarizers, the birefringence resulting from the residual strain fields could be resolved as bright and dark fringes. Very close to the microindenter, the polymer molecules are highly oriented, and the orientation dependent birefringence can be observed. This can be confirmed in materials with a spherulitic microstructure. It is observed that the initial inelastic deformation starts at very small displacement, concentrating near the corners of the indenter. As the depth of indentation increases, the deformed zone increases in size. After a depth of approximately half to

Figure 3.1 Schematic diagram showing the large-strain microindentation test: (a) front view, (b) top view

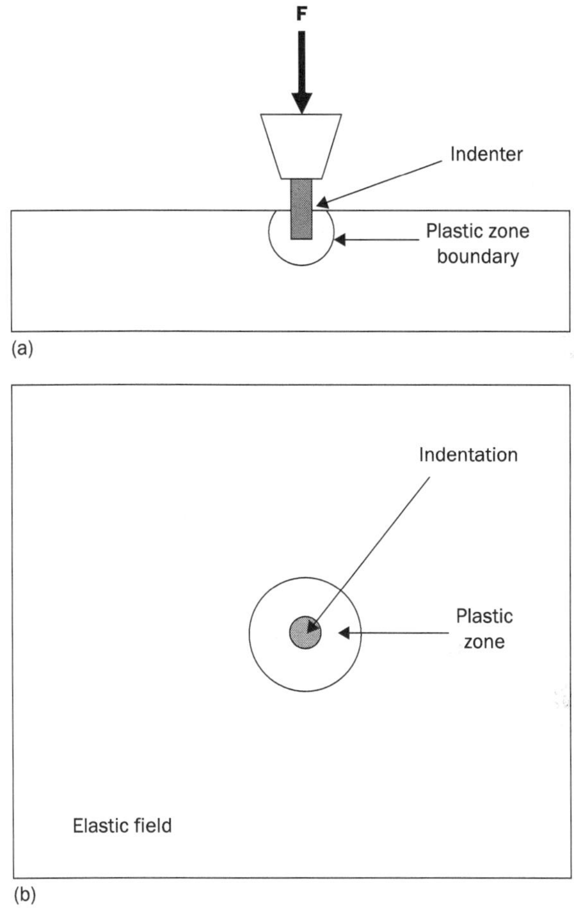

one indenter diameter, the deformation zone becomes fully developed surrounding the indenter and then remains relatively constant in size. The diameter of the deformed zone is about twice the diameter of the indenter.

The large-strain indentation under a cylindrical, flat-tipped indenter can also be simulated using the finite element method. Figure 3.3 shows the contours of the residual

Figure 3.2 Transmitted light micrographs under crossed polars showing the residual indentation deformation zone in a semi-crystalline high density polyethylene

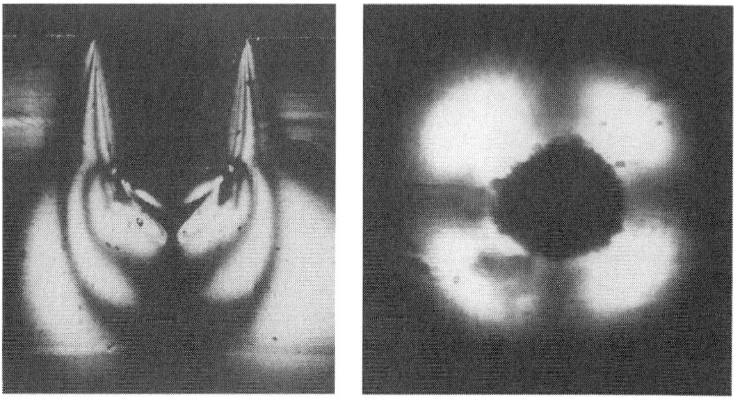

(a) (b)

Figure 3.3 Finite element simulation showing the residual indentation deformation zone in a semi-crystalline high density polyethylene

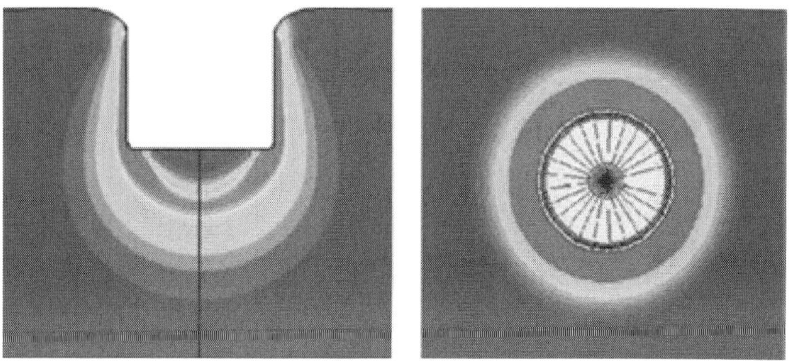

(a) (b)

equivalent plastic strain (ε^{eq}) on the HDPE, where ε^{eq} is defined by:

$$\varepsilon^{eq} = \sqrt{\frac{2}{3}(\varepsilon_1^2 + \varepsilon_2^2 + \varepsilon_3^2)} \quad [3.1]$$

and ε_1, ε_2, ε_3, are the principal strains. For $\varepsilon^{eq} > 0$, the material has yielded. Examination of the progressive development of the distribution of calculated plastic strains under the indenter shows that they are consistent with the experimental observations (Fig. 3.3). Shallow penetrations involve large plastic strains concentrating near the indenter face circumference. Later stages of penetrations (large displacements) show that the plastic zone reaches a steady state size and distribution. A transverse section through the indentation shows the distribution and size of plastic strain are similar to those experimentally observed.

Figure 3.4 shows the typical indentation stress–displacement curves using different size indenters on the HDPE sample.

Figure 3.4 Indentation stress–displacement curves of the semi-crystalline high density polyethylene for varying diameter microindenters

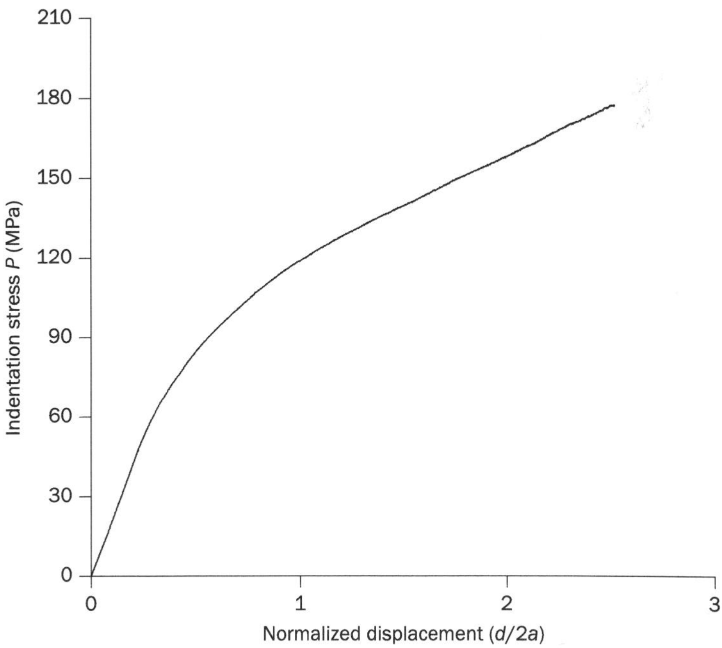

Since the contact area does not change with depth, the indentation stress (P) is simply the force applied (F) divided by the cross-sectional area of the indenter (πa^2). The deformation under the tip is clearly not uniform, and a reasonable measure of the mean indentation strain under the indenter can be obtained using the suggestion by Hill (1950) in which the penetration depth (d) is normalized with respect to the diameter ($2a$). Attempts in the present project to collapse empirically all load displacement curves onto a single curve are then successful.

The indentation stress–displacement curves reveal that the material initially deforms elastically with the applied load on the indenter, and yields at some point as the applied load is increased. The plastic deformation field and consequently the stress field progressively change with displacement, until some steady state is achieved, in which the material is continuously pushed away from the region ahead of the indenter. By using a cylindrical geometry rather than a conical or pyramidal indenter, the applied stress measured by the indenter at this steady state is constant. The achievement of this steady state can be observed in the stress–displacement curve as a distinct and sharp decrease in slope.

The large strain indentation test illustrates the inelastic deformation in the neighborhood of the indenter. The normalized stress–displacement curve could then be used to estimate the critical indentation stress (P_m), which is a measure of the yield strength of the material. When a cylindrical indenter of radius a is pressed onto a specimen, the total load (L_{total}) acting on the indenter is:

$$L_{total} = L_a + L_f \qquad [3.2]$$

where L_a is the axial load acting on the indenter end face and L_f the frictional load acting on the indenter side wall. The mean indentation pressure (P) acting on the indenter end is simply expressed as:

$$P_m = L_a/\pi a^2 \qquad [3.3]$$

The frictional load (L_f) on the indenter side wall is defined by:

$$L_f = 2\pi a h_c \tau \qquad [3.4]$$

where τ is the frictional shear stress and h_c the contact depth. Assume that the frictional stress is constant on the indenter wall, then the frictional load (L_f) should increase linearly with indentation depth, since the lateral surface area in contact with the material ($2\pi\ a\ h_c$) increases almost linearly. The critical indentation stress, P_m, is then determined by extrapolating the large-strain indentation stress–displacement curve back to zero displacement, where the frictional load (L_f) vanishes. The magnitude of P_m obtained in this way for the polyethylene studied here is approximately 66.2 MPa, from Fig. 3.4.

The critical indentation stress (P_m) is proportional to the uniaxial yield stress (σ_y) for a material. It is generally observed that the mean indentation contact pressure beneath an indenter is larger than the uniaxial compressive yield stress of the material. If yield occurs due to shear, then the mean indentation pressure required for bulk yield in an indentation test is higher than that required for a uniaxial compressive test because of the confining pressure generated by the surrounding elastically strained material in the indentation stress field. The ratio between the mean indentation pressure and the uniaxial yield stress is called the constraint factor and is given by symbol C (Tabor, 1970). For ductile metals, a value of $C \approx 3$ is generally considered to be appropriate. For soft polymers, the value of C is expected to be smaller.

Polyethylene tensile specimens have been prepared in a way similar to the material used in the indentation tests, and then deformed in uniaxial tension in a standard way. The tensile stress–strain response is seen in Fig. 3.5. The

Figure 3.5 Tensile stress–strain curve of the semi-crystalline high density polyethylene

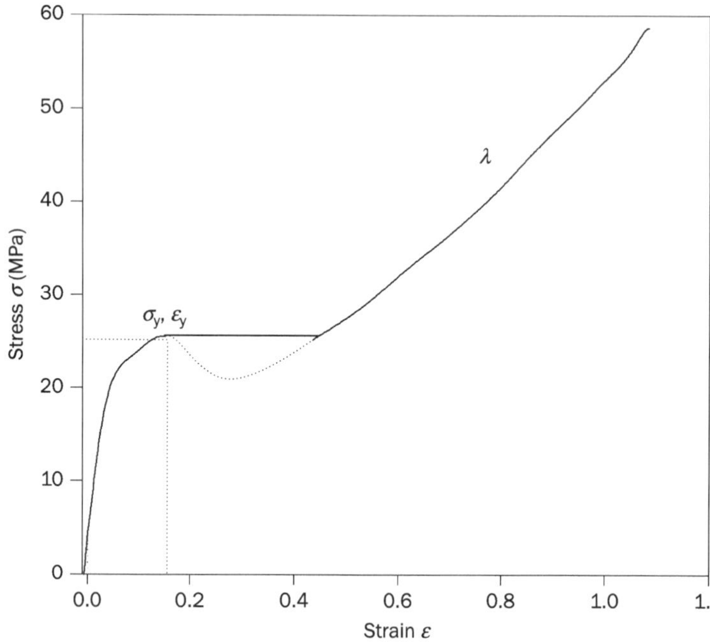

yield strength is taken as the maximum stress in the peak: σ_y = 25.4 MPa. A comparison with the indentation test shows that the critical value of the indentation stress, P_m, is approximately 2.6 times the uniaxial tensile yield stress (σ_y) for polyethylene.

3.3 Analysis of large-strain microindentation deformation

3.3.1 *Initial plastic deformation at small penetrations: the elastic analysis*

The analysis of the deformation under the indenter tip can be separated into two regions: the small displacement region and large displacement region, each of which is linear (Fig. 3.4). The 'knee' of the stress–displacement curve is related to some aspect of gross yield in the specimen. However, the micro-yield behavior of the material can be initiated at applied (indenter) stresses well below this gross yield stress. This is similar to the difference between the macroscopic bulk yield stress in metals, and the onset of the first dislocation motion in the sample, which normally occurs at a much lower stress.

The point in the indentation stress–displacement curve at which this initial microscopic yield occurs can be calculated by considering the elastic deformation field under the tip in the early stages of the penetration, well before the 'knee' of the curve is reached. The problem of a frictionless flat punch indenting an isotropic elastic half-space has been solved by Sneddon (1946) (with misprints later corrected by Barquins and Maugis, 1982). The stresses produced in the semi-infinite elastic solid by the rigid flat-ended punch of radius a were analyzed. For the present study, the stress distributions in a semi-infinite HDPE block pressed by the flat punch were calculated based on Sneddon's approach as follows.

The indentation of a semi-infinite solid by a rigid flat punch was axially symmetric. Cylindrical polar coordinates z, r, θ are used (Fig. 3.6). It was assumed that the z-component of the surface displacement was specified over the region $r < a$ of the surface $z = 0$, while outside that region the condition of zero normal stress was imposed. It was further assumed that the tangential stress, τ_{zr}, vanished at all points of the boundary. Thus for $z = 0$ the boundary conditions were:

$$
\begin{aligned}
&u_z = u_z(r),\ 0 \le r \le a \\
&\sigma_z = 0,\ r > a \qquad [3.5] \\
&\tau_{zr} = 0,\ \text{for all values of } r
\end{aligned}
$$

Figure 3.6 Coordinate system for axially symmetric flat indenter loading

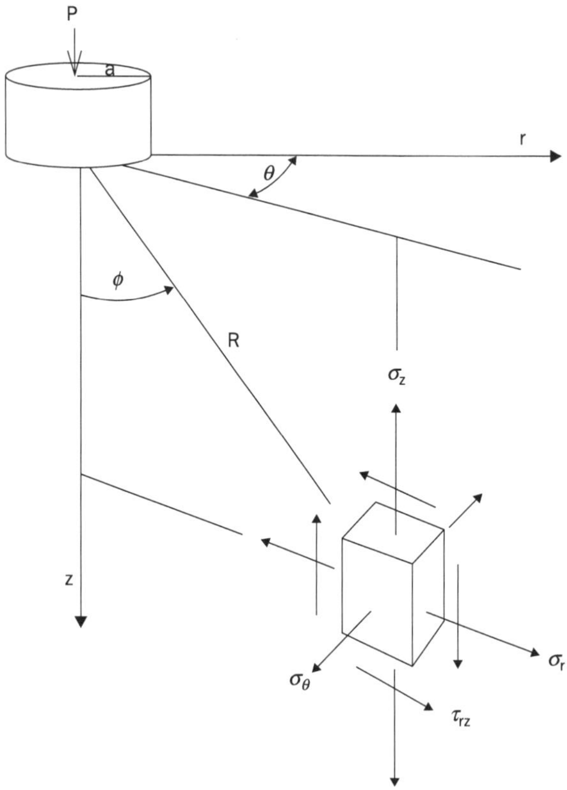

The stress components (σ_r, σ_θ, σ_z, τ_{rz}), under a compressive indentation pressure P, were then given as follows:

$$\sigma_z = -\frac{1}{2}\left[J_1^0 + \frac{z}{a}\cdot J_2^0\right] \qquad [3.6]$$

$$\sigma_r = -\frac{1}{2}\left[J_1^0 - \frac{z}{a}\cdot J_2^0 - (1-2v)\cdot\frac{a}{r}\cdot J_0^1 + \frac{z}{r}\cdot J_1^1\right] \qquad [3.7]$$

$$\sigma_\theta = -\frac{1}{2}\left[2vJ_1^0 + (1-2v)\cdot\frac{a}{r}\cdot J_0^1 - \frac{z}{r}\cdot J_1^1\right] \qquad [3.8]$$

$$\tau_{rz} = -\frac{1}{2}\frac{z}{a} \cdot J_2^1 \qquad [3.9]$$

with the following notations:

$$J_1^0 = R^{-1/2} \cdot \sin\frac{\varphi}{2} \qquad [3.10]$$

$$J_0^1 = \frac{a}{r} \cdot \left(1 - R^{1/2} \cdot \sin\frac{\varphi}{2}\right) \qquad [3.11]$$

$$J_2^1 = \frac{r}{a} \cdot R^{-3/2} \cdot \sin\frac{3\varphi}{2} \qquad [3.12]$$

$$J_2^0 = \left[1 + \frac{z^2}{a^2}\right]^{1/2} \cdot R^{-3/2} \cdot \sin\left(\frac{3\varphi}{2} - \theta\right) \qquad [3.13]$$

$$J_1^1 = \left[1 + \frac{z^2}{a^2}\right]^{1/2} \cdot \frac{a}{r} \cdot R^{-1/2} \cdot \sin\left(\theta - \frac{\varphi}{2}\right) \qquad [3.14]$$

$$R = \left\{\left[\frac{r^2}{a^2} + \frac{z^2}{a^2} - 1\right]^2 + 4 \cdot \frac{z^2}{a^2}\right\}^{1/2} \qquad [3.15]$$

$$\tan\varphi = 2 \cdot \frac{z}{a} \cdot \frac{1}{(r^2/a^2) + (z^2/a^2) - 1} \qquad [3.16]$$

$$\tan\theta = a/z \qquad [3.17]$$

where r is the radial distance, z the distance into the specimen, a the indenter radius and v the Poisson's ratio. For HDPE, the Poisson's ratio was chosen as $v = 0.3$ (Young, 1991). Other variables were calculated as long as r, z, a, v are known.

The principal stresses at any point were given as:

$$\sigma_1 = \frac{\sigma_r + \sigma_z}{2} + \left[\left(\frac{\sigma_r - \sigma_z}{2}\right)^2 + \tau_{rz}^2\right]^{1/2} \qquad [3.18]$$

$$\sigma_2 = \sigma_\theta \qquad [3.19]$$

$$\sigma_3 = \frac{\sigma_r + \sigma_z}{2} - \left[\left(\frac{\sigma_r - \sigma_z}{2}\right)^2 + \tau_{rz}^2\right]^{1/2} \qquad [3.20]$$

where $\sigma_1 \geq \sigma_2 \geq \sigma_3$. σ_1 and σ_3 are the principal normal stresses, contained in the symmetry plane θ = constant; σ_2 is everywhere perpendicular to the symmetry plane, and is defined as the 'hoop' stress.

The load at which plastic yield begins in the stress field of two solids in contact is related to the yield stress of the softer material. Following Hill (1950), the octahedral shear stress is a defined function of the octahedral shear strain for all strain configurations; the principal axes of stress and strain in a given element coincide and do not rotate during plastic flow. The octahedral shear stress, τ_n, is defined by the equation:

$$\tau_n = \frac{1}{3}\{(\sigma_2 - \sigma_3)^2 + (\sigma_3 - \sigma_1)^2 + (\sigma_1 - \sigma_3)^2\}^{\frac{1}{2}} \qquad [3.21]$$

where σ_1, σ_2, σ_3 are the principal stresses. The yield of many ductile materials is thus described by von Mises' shear strain–energy criterion:

$$\tau_n = \sqrt{2}\sigma_y/3 \qquad [3.22]$$

where σ_y donates the value of the yield stress of material in simple tension or compression.

The contour plots of principal stresses σ_1, σ_2, σ_3 with respect to the radial distance (r/a) and vertical distance (z/a) are presented in Figs. 3.7–3.9. The octahedral shear stress was then calculated using Eq. [3.21], and is plotted in Fig. 3.10. The stress at a point is proportional to the nominal applied punch stress (P) and is normalized with respect to this applied stress.

As the applied stress P increases in the small strain region, the maximum value of the octahedral stress is approximately

Figure 3.7 Contour of principal stress σ_1 under the punch, following Sneddon (1946)

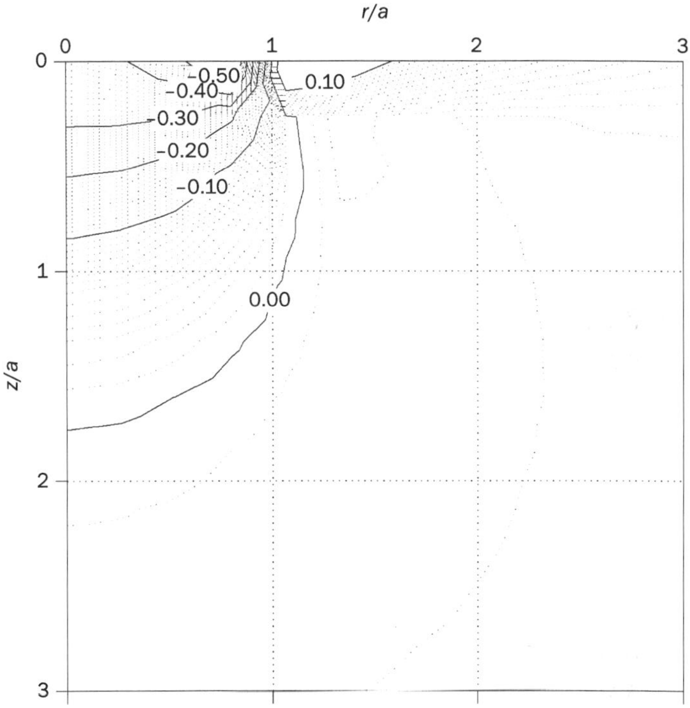

0.38 *P* at the point labeled 'A' in Fig. 3.10. This is the first region to yield following the small volume of plastic deformation which occurs at the corner of the indenter. Young (1991) has shown that the yield stress for polyethylene is pressure dependent. The material under the indenter is constrained, and the hydrostatic component of the stress field is shown in Fig. 3.11. For a material at the yield point at 'A', the hydrostatic stress at 'A' is much smaller than the yield stress. This is not sufficient to induce any significant increase in yield stress locally. Hence the pressure independent von Mises yield criterion is a reasonable approximation. The maximum Mises shear stress is 0.38 *P* at the point 'A'. Substituting in Eq. [3.22], we have:

Figure 3.8 Contour of principal stress σ_2 under the indenter, following Sneddon (1946)

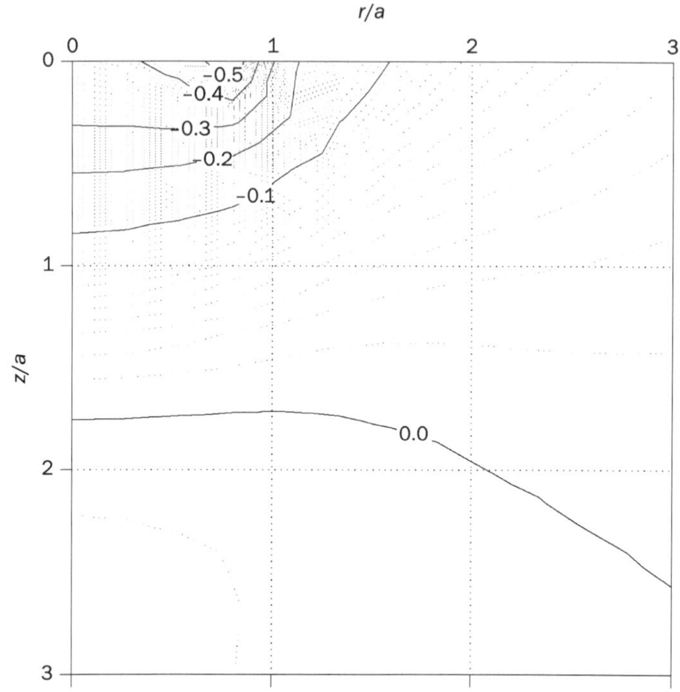

$$(0.38 \times P) = \tau_n = \frac{\sqrt{2}}{3}\sigma_y \qquad [3.23]$$

and thus

$$\frac{P_{onset}}{\sigma_y} = 1.25 \qquad [3.24]$$

where P_{onset} is the measured indentation stress at which yielding is initiated at the point 'A' in the specimen, and σ_y is the tensile yield stress of HDPE. This initial yielding corresponds to the early stages of penetration, before the plastic zone has grown to its steady state size and occurs well

Figure 3.9 Contour of principal stress σ_3 under the indenter, following Sneddon (1946)

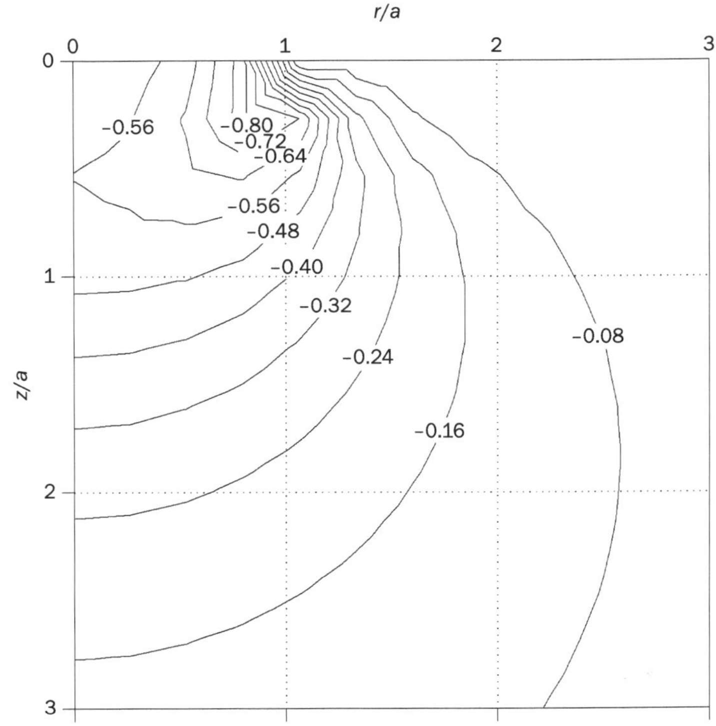

before the gross yield point (at the 'knee' of the load–displacement curve).

The deformation under the tip therefore involves progressively increasing applied stress P, and localized yield at the points near the indenter corners occurs at an applied indenter stress $P = 1.25\ \sigma_y$, which is still well within the first linear region of the indentation curve (Fig. 3.4). The curve remains visibly linear beyond this depth, indicating the plastic volume is relatively small until the later 'knee' to the curve is observed. Otherwise the micro-yield point is undetectable from the monotonic stress–displacement curve.

Figure 3.10 Octahedral shear stress (τ_n) field (normalized with respect to the applied stress) under an indenter. The maximum of τ_n is found near the point A

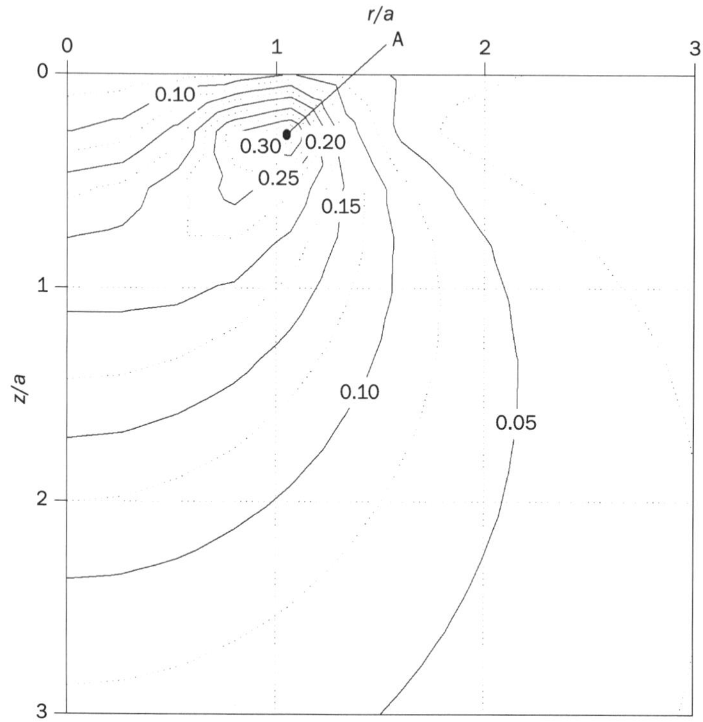

3.3.2 Full plastic deformation at large penetrations: the cavitation models

In the deep penetration test the critical indentation stress (P_m) is measured at and after yield. In this part of the test, where there is considerable plastic deformation in the material, the indentation process can be studied using variations on earlier cavitation models. The penetration of the indenter is modeled as the opening of a cavity in the material. The stresses to open such a cavity and the related

Figure 3.11 Hydrostatic stress (normalized with respect to the applied stress) under an indenter

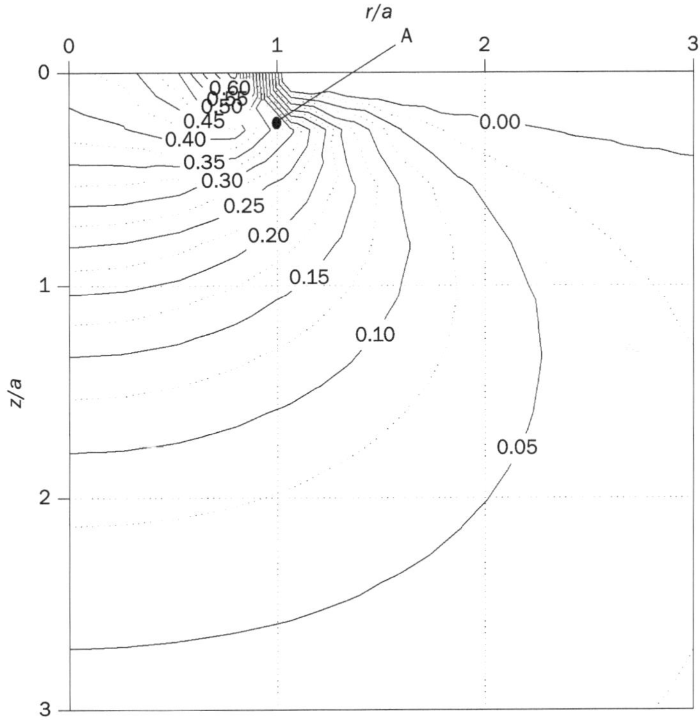

indentation stress required can be estimated. This cavity is assumed to start with an initial radius a_o, and opens to a final radius a, equal to the radius of the indenter. The opening of such a cavity also expands a surrounding plastic zone from an initial radius r_o to a final radius c (Fig. 3.12). The stress applied to the indenter to cause it to penetrate the material can be calculated by calculating the pressure to open this hypothetical cavity. The material properties which oppose this cavity opening can then be inferred from the calculation.

As seen in Fig. 3.5, the crystalline polymers normally have very different stress–strain behavior from that of the metals. Typical characteristics include: (a) an elastic strain range of

Figure 3.12 Schematic diagram showing the expansion of a hypothetical cavity opening under a cylindrical indenter

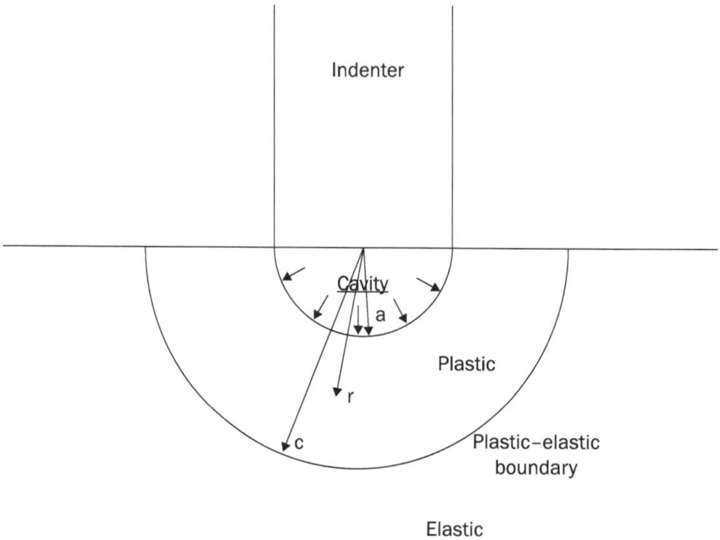

the order of 0.1, very large compared with metals; (b) a significant departure from linearity in this pre-yield range; (c) a rather broad maximum in the stress–strain curve; (d) a range, following the maximum, of strain softening, etc. In order to obtain the proper constitutive parameters, an idealization for the experimental stress–strain curve is often needed. The idealized stress–strain curve of polyethylene is obtained based on the original experimental result (Fig. 3.5). The values corresponding to the maximum stress right after the initial yielding are the yield stress σ_y and yield strain ε_y. The strain softening region has been ignored and the post-yield region is assumed to be in exponential form and described by the strain hardening coefficient λ. The constitutive relationship between the true stress σ and the logarithmic strain ε for crystalline polymers can be described by (Wright et al., 1992):

$$\sigma/\sigma_y = f(\varepsilon) = \begin{cases} \varepsilon/\varepsilon_y & \varepsilon \le \varepsilon_y \\ \exp(\lambda \cdot (\varepsilon - \varepsilon_y)) & \varepsilon > \varepsilon_y \end{cases} \quad [3.25]$$

where λ is the exponential hardening index, and ε_y is the strain at yield.

There are two approaches for opening a cavity, as suggested by Bishop, Hill and Mott (1945):

1. Cylindrical cavitation: starting with a cylindrical hole of infinite length, to find the pressure P_m, which will enlarge the hole.
2. Spherical cavitation: starting with a small hollow sphere in the body of an infinite ductile material capable of work-hardening, to find the pressure P_m.

Cylindrical cavitation model

Following Bishop et al. (1945), the octahedral shear stress (τ_n) and the corresponding octahedral shear strain (γ_n) are used to define the deformation behavior in a multiaxial loading. τ_n and γ_n are proportional to the true stress (σ) and strain (ε) in a uniaxial tension or compression test:

$$\tau_n = \sqrt{2}\sigma/3, \ \gamma_n = \sqrt{2}\varepsilon \quad [3.26]$$

By adopting Eq. [3.26], Eq. [3.25] becomes:

$$\frac{(3/\sqrt{2}) \cdot \tau_n}{\sigma_y} = f\left(\frac{\gamma_n}{\sqrt{2}}\right) \quad [3.27]$$

Following Bishop et al. (1945), let σ_r, σ_θ be the radial and tangential stresses in the cylindrical polar coordinates (z, r, θ), and the octahedral shear stress (τ_n) and octahedral shear strain (γ_n) become:

$$\tau_n = \frac{1}{3} \cdot \sqrt{\frac{3}{2}} \cdot (\sigma_\theta - \sigma_r),\ \gamma_n = 2\sqrt{\frac{2}{3}}|\varepsilon_r| \qquad [3.28]$$

It therefore follows from Eq. [3.27] that the stress–strain relation is:

$$\frac{\sigma_\theta - \sigma_r}{\sigma_y} = \frac{2}{\sqrt{3}} \cdot f\left(\frac{2}{\sqrt{3}}|\varepsilon_r|\right) \qquad [3.29]$$

Integrating the condition of the equilibrium, namely:

$$\frac{d\sigma_r}{dr} = \frac{\sigma_\theta - \sigma_r}{r} \qquad [3.30]$$

throughout the plastic region from a to r, we obtain for the pressure P on the boundary of the hollow cylinder (the value of $(-\sigma_r)$ at that point):

$$\frac{P}{\sigma_y} = \frac{2}{\sqrt{3}} \cdot \int_a^r f\left\{\frac{2}{\sqrt{3}}|\varepsilon_r|\right\}\frac{dr}{r} \qquad [3.31]$$

To evaluate the integral, an expression for ε_r is found next. Using the same type of approach used by others (Bishop et al., 1945; Wright et al., 1992), if an element at radius r in the plastic region when the hole is of radius a was formerly at radius r_o when the hole was of zero radius, then:

$$r^2 - a^2 = r_o^2 - a_o^2 \qquad [3.32]$$

and

$$-\varepsilon_r = \varepsilon_\theta = -\ln\left(\frac{r}{r_o}\right) = -\frac{1}{2} \cdot \ln\left(1 - \frac{a^2 - a_o^2}{r^2}\right) \qquad [3.33]$$

thus Eq. [3.31] becomes:

$$\frac{P}{\sigma_y} = \frac{2}{\sqrt{3}} \cdot \int_a^r f\left\{-\frac{1}{\sqrt{3}}\ln\left(1 - \frac{a^2 - a_o^2}{r^2}\right)\right\}\frac{dr}{r} \qquad [3.34]$$

The above finite strain analysis indicates the existence of a cavitation limit P_m, as $a/a_o \to \infty$

$$\frac{P_m}{\sigma_y} = \frac{2}{\sqrt{3}} \cdot \int_1^{\eta} f\left\{-\frac{1}{\sqrt{3}}\ln\left[1-\frac{1-(a_o/a)^2}{\eta^2}\right]\right\}\frac{d\eta}{\eta} \qquad [3.35]$$

where $\eta = r/a$, and the value of η is found via Eq. [3.25] and Eq. [3.35]:

$$\eta = \left[1-\left(\frac{a_o}{a}\right)^2\right]^{\frac{1}{2}} \cdot [1-\exp(-\sqrt{3}\varepsilon)]^{-\frac{1}{2}} \qquad [3.36]$$

This cylindrical solution is found to be identical to that developed by Wright et al. (1992).

Spherical cavitation model

In the expansion of a spherical hole, the octahedral stress and strain are related to the radial and tangential stresses (Bishop et al., 1945):

$$\tau_n = \frac{\sqrt{2}}{3} \cdot (\sigma_\theta - \sigma_r),\ \gamma_n = 2\sqrt{2}|\varepsilon_r| \qquad [3.37]$$

It therefore follows from Eq. 3.25 that the stress–strain relation is:

$$\frac{\sigma_\theta - \sigma_r}{\sigma_y} = f(2|\varepsilon_r|) \qquad [3.38]$$

Integrating the condition of the equilibrium (Bishop et al., 1945), namely:

$$\frac{d\sigma_r}{dr} = \frac{2 \cdot (\sigma_\theta - \sigma_r)}{r} \qquad [3.39]$$

throughout the plastic region from a to r, we obtain for the pressure P on the boundary of the hollow sphere hole (the value of $(-\sigma_r)$ at that point):

$$\frac{P}{\sigma_y} = 2 \cdot \int_a^r f\{2|\varepsilon_r|\}\frac{dr}{r} \qquad [3.40]$$

Similarly, an expression for ε_r is needed to evaluate the integral. In the case of a spherical hole,

$$r^3 - a^3 = r_o^3 - a_o^3 \qquad [3.41]$$

and

$$-\varepsilon_r = \varepsilon_\theta = -\ln\left(\frac{r}{r_o}\right) = -\frac{1}{3}\cdot\ln\left(1 - \frac{a^3 - a_o^3}{r^3}\right) \qquad [3.42]$$

thus, Eq. [3.40] becomes:

$$\frac{P}{\sigma_y} = 2\cdot\int_a^r f\left\{-\frac{2}{3}\ln\left(1 - \frac{a^3 - a_o^3}{r^3}\right)\right\}\frac{dr}{r} \qquad [3.43]$$

The above finite strain analysis indicates the existence of a cavity limit P_m, as $a/a_o \to \infty$

$$\frac{P_m}{\sigma_y} = 2\cdot\int_1^\eta f\left\{-\frac{2}{3}\ln\left[1 - \frac{1-(a_o/a)^3}{\eta^3}\right]\right\}\frac{d\eta}{\eta} \qquad [3.44]$$

here $\eta = r/a$, ranging from 1 to ∞.

The value of η is found, via Eq. [3.25] and Eq. [3.44]:

$$\eta = \left[1 - \left(\frac{a_o}{a}\right)^3\right]^{\frac{1}{3}} \cdot \left[1 - \exp\left(-\frac{3}{2}\varepsilon\right)\right]^{-\frac{1}{3}} \qquad [3.45]$$

The critical indentation pressure P_m can thus be estimated with the help of cavitation models, and the solution can be either cylindrical (Eq. [3.35]) or spherical (Eq. [3.44]). The cavitation pressure P predicted from the spherical model is compared with the cylindrical cavitation model, as seen in Fig. 3.13. As a_o decreases, the ratio P/σ_y for both models approaches a constant value corresponding to P_m/σ_y, where P_m is the critical indentation pressure. This stress also varies with work hardening characteristics (λ). The hardening index $\lambda = 1.2$ is used in the present calculation, a typical value for polymers. It is seen that there is a significant difference between the spherical and cylindrical cavitation model predictions.

Figure 3.13 Comparison of the spherical (solid line) and cylindrical (dotted line) cavitation model predictions for the critical indentation stress. Calculated for typical yield strain of ε_y = 0.1 and strain hardening index of λ = 1.2

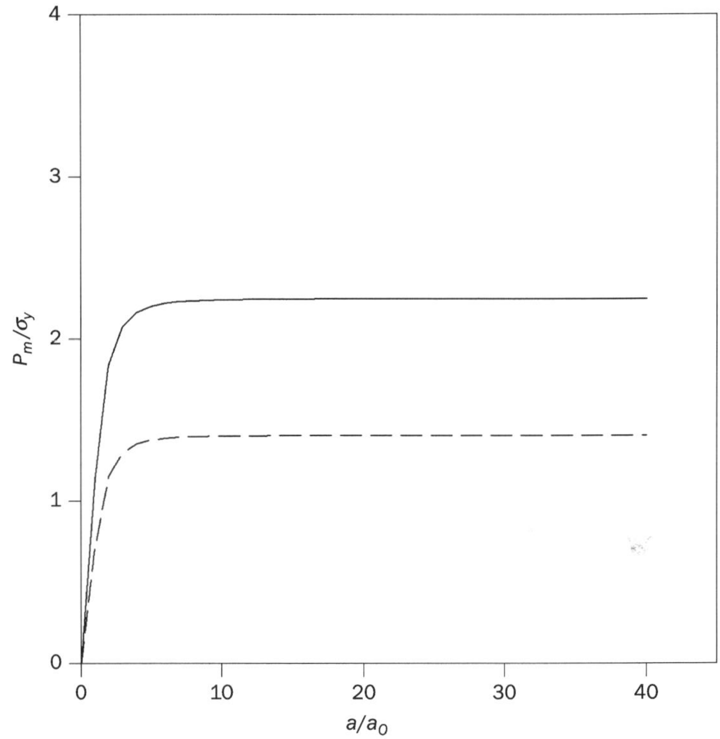

3.4 Estimates of plastic properties of crystalline polymers by large-strain indentation

3.4.1 Critical indentation stress

Experimentally, since the critical value of the indentation stress can be measured, the model predictions (Eq. [3.35] and [3.44]) can be more usefully plotted in the following way. In the limit for small a_o ($a/a_o \rightarrow \infty$), for the spherical cavitation model:

$$P_m^s = \sigma_y \cdot \int_1^{\eta} \frac{\exp[\lambda \cdot (\varepsilon - \varepsilon_y)]}{\exp\left(\frac{3}{2}\varepsilon\right)} d\varepsilon \qquad [3.46]$$

where $\eta = \left[1 - \exp(-\frac{3}{2}\varepsilon)\right]^{-\frac{1}{3}}$, and for the cylindrical cavity model:

$$P_m^c = \sigma_y \cdot \int_1^{\eta} \frac{\exp[\lambda \cdot (\varepsilon - \varepsilon_y)]}{\exp(\sqrt{3}\varepsilon)} d\varepsilon \qquad [3.47]$$

where $\eta = [1 - \exp(-\sqrt{3}\varepsilon)]^{-\frac{1}{2}}$.

It is clear that the forms of Eq. [3.46] and [3.47] are similar to the broadly applicable empirical relationship suggested by Tabor (1986 and 1996):

$$P_m = \sigma_y \cdot C \qquad [3.48]$$

However, the physical significance of this similarity is not clear. The term C is a 'constraint' factor, the magnitude of which depends upon the material properties, as indicated by the ratio of E/σ_y (Tabor, 1986 and 1996; Johnson, 1970; Fleck et al., 1992; Fischer-Cripps, 1997). For most metals, the ratio E/σ_y is greater than 100, which in turn predicts a value of $C \approx 3$; for polymers, for which E/σ_y is of the order of 10, a lower constraint factor is expected (Tabor, 1986).

The critical value of the indentation stress (P_m) shown in Eq. [3.46] and [3.47] varies with the yield strength, the yield strain and the work hardening behavior. For typical polymeric materials, the model predictions can be seen in Fig. 3.14. The 'constraint factors' predicted for polymeric materials are observed in the right range ($C \approx 3$). Again, the predictions show the significant difference between the two models.

The estimation of the critical indentation stress P_m requires the constitutive parameters (σ_y, ε_y, λ) as input (Eq. [3.35] and

Figure 3.14 Comparison of the critical indentation stress predicted from the spherical (solid line) and cylindrical (dotted line) cavitation models for typical variations of plastic properties for polymers. Calculated for a typical yield strain of $\varepsilon_y = 0.1$

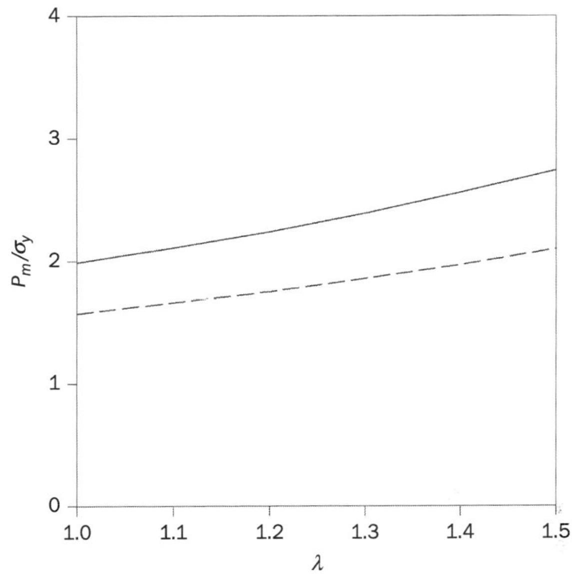

[3.44]). These parameters can be obtained from experimental stress–strain measurements. The tensile stress–strain curve of polyethylene has been given earlier (Fig. 3.5). The values corresponding to the maximum stress right after the initial yielding are the yield stress σ_y and yield strain ε_y. The strain softening region has been ignored and the strain hardening coefficient λ is estimated by curve-fitting the stress–strain curve after the flat plateau region. The results of the constitutive parameters of slow-cooled polyethylene are given in Table 3.1.

By inputting the constitutive parameters into Eq. [3.46] and [3.47], the critical indentation stress (P_m) for polyethylene is calculated, as seen in Fig. 3.15. This slow-cooled polyethylene

Table 3.1 Constitutive parameters of HDPE determined from the tensile stress–strain curve

Specimen	Tensile yield stress σ_y *(MPa)*	Tensile yield strain ε_y	Exponential hardening index λ
HDPE (crystallinity = 76%)	25.4	0.11	1.23

Figure 3.15 A comparison of the critical indentation stress P_m between the experimental results and the cavitation models (spherical – solid line, cylindrical – dotted line) for the semi-crystalline high density polyethylene

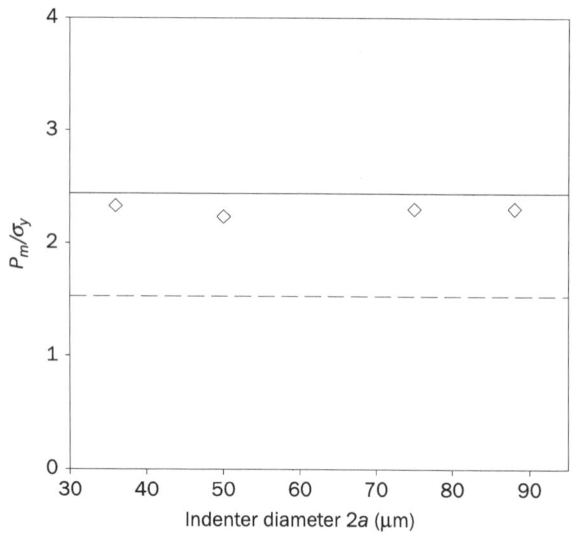

has been experimentally examined using varying sizes of microindenter ($2a$ = 36 µm, 50 µm, 68 µm, 88 µm). From the indentation stress–displacement curves, the critical indentation stresses are obtained by extrapolating the large-strain stresses back to zero displacement. The results are included in Fig. 3.15 (dotted points). In comparison, it is found that the spherical model prediction fits more closely to the experimental data.

3.4.2 Plastic zone size

In the cavitation state the plastic zone radius c increases linearly with the void radius a in the deformed configuration. At the edge of the plastic zone, $\varepsilon = \varepsilon_y$, Eq. [3.36] and [3.45] give in the limit of $a/a_o \to \infty$,

$$\frac{c^s}{a} = \left[1 - \exp\left(-\frac{3}{2}\cdot\varepsilon_y\right)\right]^{-\frac{1}{3}} \quad [3.49]$$

for the spherical cavity model, and

$$\frac{c^c}{a} = \left[1 - \exp(-\sqrt{3}\cdot\varepsilon_y)\right]^{-\frac{1}{2}} \quad [3.50]$$

for the cylindrical cavity model.

In both cases, c represents the radius of the plastic–elastic boundary. The extent of the plastic zone therefore depends on the yield strain of the material, and is different for the two models. For polyethylene, the typical yield strains are near 0.10 (from Fig. 3.5). The plastic zone sizes were measured from photographs taken through the polished side wall (Fig. 3.2) and the finite element simulation (Fig. 3.3). The results are summarized in Table 3.2. Consistent with the earlier observations, the spherical cavitation model is a better fit to the experimental observations.

Table 3.2 **Summary of plastic zone sizes of the high density polyethylene indented with a cylindrical, flat-ended tip**

	Plastic zone size c/a
Spherical cavitation model	1.96
Experimental measurement (36 μm tip)	2.03
Experimental measurement (88 μm tip)	1.96
Finite element simulation	2.10
Cylindrical cavitation model	2.50

3.4.3 Effect of tip geometry

The present study has used the cylindrical indenter equipped with a flat-ended tip. The tips used for micro-/nanoindentation tests are typically prepared by an electropolishing process. It is always challenging to prepare a perfectly flat tip during the polishing process. Intuitively, it might be expected that the exact shapes at the indenter tips may affect the deformation of the indentation and resultant load–displacement responses. The strain fields for a cylindrical indenter with different geometry tips have been calculated using the finite element method, as shown in Fig. 3.16. It is seen that, at large

Figure 3.16 Contours of equivalent plastic strain at large-displacement indentation under cylindrical indenters of various tip geometry: (left) flat tip, (middle) conical tip and (right) spherical tip

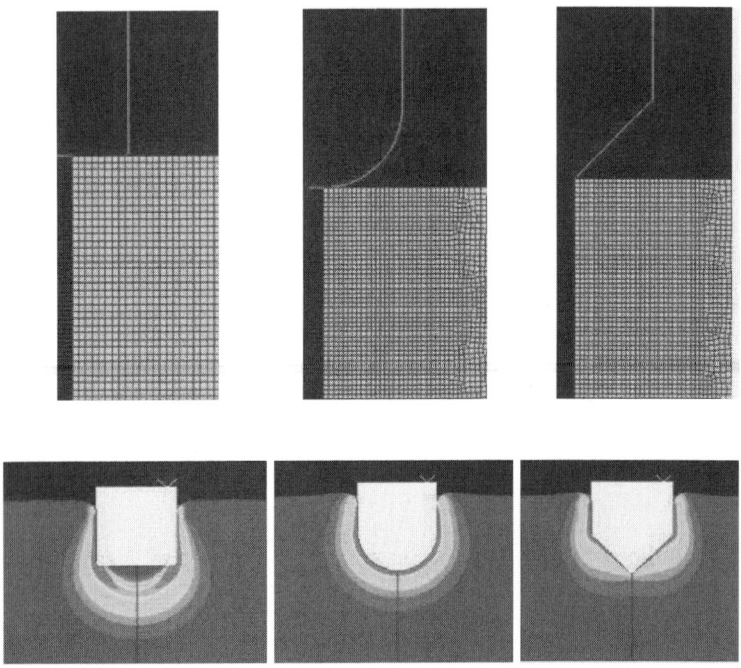

displacement, the penetration of the indenter is accommodated by radial displacement of the material regardless of the tip geometry. The deformation near the edges of the tips might differ slightly, but the radial extent of the plastic zone at large displacement is approximately the same, about twice the indenter radius.

3.5 Conclusions

Large-strain microindentation has been performed on a semi-crystalline high-density polyethylene using a cylindrical, flat-ended microindenter. The plastic deformation under the tip was studied experimentally and analyzed. The stress–displacement curve which was the experimental output of the test was modeled using the cavitation approach. Material properties such as yield stress and strain and work hardening characteristics could be input into the model and the variations in measured indentation stress–displacement curves predicted accurately. Microindentation provides a reliable method to measure the plastic properties polymers in small volumes.

3.6 References

ABAQUS (2009), *ABAQUS Theory Manual V.6.8*. Pawtucket: ABAQUS Inc.

Anton, R.J., Miskiohlu, I. and Subhash, G. (1999), 'Determination of residual stress fields beneath a Vickers indentation using photoelasticity', *Experimental Mechanics*, 39: 227.

Barquins, M. and Maugis, D. (1982), 'Adhesive contact of axisymmetric punches on an elastic half-space: the modified Hertz–Hubers stress tensor for contacting spheres', *J. Mec. Theo. Appl.*, **1**: 331–57.

Bishop, R.E., Hill, R. and Mott, N.F. (1945), 'The theory of indentation and hardness tests', *Proceeding of Physics Society*, **57**: 147.

Doerner, M.F. and Nix, W.D. (1986), 'A method for interpreting the data from depth-sensing indentation instruments', *Journal of Materials Research*, **1**: 601–9.

Fischer-Cripps, A.C. (1997), 'Elastic–plastic behavior in materials loaded with a spherical indenter', *Journal of Materials Science*, **32**: 727–36.

Fleck, N.A., Otoyo, H. and Needleman, A. (1992), 'Indentation of porous solid', *International Journal of Solids and Structures*, **29**: 1613–36.

Gubicza, J., Luhasz, A. and Tasnadi, P. (1996), 'Determination of the hardness and elastic modulus from continuous Vickers indentation testing', *Journal of Materials Science – V*. **31**: 3109–14.

Herbert, E.G., Pharr, G.M., Oliver, W.C., Lucas, B.N. and Hay, J.L. (2001), 'On the measurement of stress–strain curves by spherical indentation', *Thin Solid Films*, **331–5**: 398–9.

Hill, R. (1950), *The Mathematical Theory of Plasticity*, Oxford: Oxford University Press.

Jayaraman, S., Hahn, G.T., Oliver, W.C. and Rubin, C.A. (1998), 'Determination of monotonic stress–strain curve of hard materials from ultra-low-load indentation tests', *International Journal of Solids and Structures*, **35**: 365–81.

Jervis, T. R. and Spaepen, F. (1994), 'Nanoindentation of Ag/Ni multilayered thin films', *Journal of Applied Physics – V*, 75, 4969–74.

Johnson, K.L. (1985) *Contact Mechanics*, Cambridge: Cambridge University Press.

Loubet, J.L., Georges, J.M., Marchesini, J.M. and Meilie, G. (1984), 'Vickers indentation curves of magnesium oxide (MgO)', *Journal of Tribology*, **106**: 43.

Lu, Y.C., Kurapati, S. K. N. and Yang, F. (2008), 'Finite element analysis of deep indentation by a spherical indenter', *Journal of Materials Science*, **43**(18): 6331–6.

Lu, Y.C. and Shinozaki, D.M. (2010), 'Temperature dependent viscoelastic properties of polymers investigated by small-scale dynamic mechanical analysis', *Experimental Mechanics*, **50**: 71–7.

Lu, Y.C., Tandon, G. P., Jones, D.C. and Schoeppner, G. A. (2009), 'Elastic and viscoelastic characterization of thermally oxidized polymer resin using nanoindentation', *Journal of Mechanics of Time-Dependent Materials*, **13**: 245–60.

Lu, Y.C., Tandon, G. P., Putthanarat, S. and Schoeppner, G.A. (2009), 'Nanoindentation strain rate sensitivity of PMR-15 Polyimide', *Journal of Materials Science*, **44**: 2119–27 (2009).

Marshall, D. B., Evans, A. G. and Rossington, C. (1984), 'Measurement of adherence of residually stressed thin films by indentation: mechanics of interface delamination', *Journal of Applied Physics – V*, **56**: 2632–44.

Marshall, D. B. (1984), 'An indentation method for measuring matrix-fiber frictional stresses in ceramic composites', *Journal of the American Ceramic Society* **67**: C259–60.

Mayo, M.J. and Nix, W.D. (1988), 'A microindentation study of superplasticity in Pb, Sn, and Sn-Wt38% Pb', *Acta Metallurgy and Materials*, **36**(8): 2183–92.

O'Neill, H. (1967), *Hardness Measurement of Metals and Alloys*, London: Chapman and Hall Ltd.

Ogilvy, J. A. (1993), 'Predicting the friction and durability of MoS_2 coatings using a numerical contact model', *Wear*, **160**: 171–80.

Oliver, W.C., Hutchings, R. and Pethica, J.B. (1986), in B.J. Blau and B.R.Lawn (Eds), *Microindentation Techniques in Materials Science and Engineering*, Pennsylvania: ASTM.

Oliver, W. C. and Pharr, G. M. (1992), 'An Improved Technique for Determining Hardness and Elastic Modulus Using Load and Displacement Sensing Indentation Experiments', *Journal of Material Research*, **7**: 1564–83.

Park, Y.J. and Pharr, G.M. (2004), 'Nanoindentation with spherical indenters: finite element studies of deformation in the elastic–plastic transition regime', *Thin Solid Films*, **447–8**: 246–50.

Pharr, G.M., Harding, D.S. and Oliver, W.C. (1994), 'Measurement of Fracture Toughness in Thin Films and Small Volumes Using Nanoindentation Methods', *Materials Research Society Symposium, 1994.*

Shinozaki, D.M. and Lu, Y. (1997), 'Micro-indentation relaxation measurements in polymer thin films', *Journal of Electronic Materials*, **26**: 852.

Shinozaki, D.M. and Klauzner, A. (1991), 'Stress relaxation of high impact polystyrene', *Journal of Materials Science*, **26**: 5865–72.

Sneddon, I.N. (1946), 'Boussinesq's problem for a flat-ended cylinder', *Proceeding of Cambridge Philosophy Society*, **42**: 29.

Swadener, J.G., Taljat, B. and Pharr, G.M. (2001), 'Measurement of residual stress by load and depth sensing indentation with spherical indenters', *Journal of Material Research*, **16**: 2091.

Swain, M.V. (1998), 'Mechanical property characterisation of small volumes of brittle materials with spherical tipped indenters', *Materials Science and Engineering A*, **253**: 160–6.

Tabor, D. (1986), 'Indentation hardness and its measurement: some cautionary comments', In B.J. Blau and B.R. Lawn

(Eds), *Microindentation Techniques in Materials Science and Engineering*, Pennsylvania: ASTM.

Tabor, D. (1996), 'Indentation hardness: fifty years on a personal view', *Philosophical Magazine*, **A74**: 1207.

Tabor, D. (1970), 'The hardness of solids', *Review of Physics in Technology*, p. 145.

Vlassak, J.J. and Nix, W.D. (1994), 'Measuring the elastic properties of anisotropic materials by means of indentation experiments', *Journal of Mechanics and Physics of Solids*, **42**: 1223.

Vlassak, J.J. and Nix, W.D. (1993), 'Indentation modulus of elastically anisotropic half spaces', *Philosophical Magazine*, **A67**: 1045.

Wright, S.C., Huang, Y. and Fleck, N.A. (1992), 'Deep penetration of polycarbonate by a cylindrical punch', *Mechanics of Materials*, **13**: 277.

Young, R.J. (1991), *Introduction to Polymers*, New York: Chapman and Hall.

Zhang, L. and Mahdi, M. (1996), 'The plastic behaviour of silicon subjected to micro-indentation', *Journal of Material Science*, **31**: 5671–76.

4

Nanocomposite coatings: A review

L. Santo, University of Rome 'Tor Vergata', Italy, and J. P. Davim, University of Aveiro, Portugal

Abstract: This chapter discusses nanocomposite coatings and provides an overview of different nanocomposite coatings applied on various substrates. It describes the main deposition techniques, the coating properties and characterization methods; gives a brief description of nanocomposite polymer coating, and discusses some possible applications in areas such as the automotive, aerospace, electronics and biomedical sectors. Finally, perspectives for the field and future trends are reported.

Key words: nanocomposite coatings, hard and superhard coatings, mechanical and tribological properties.

4.1 What are nanocomposite coatings?

Nanocomposite coatings are a new class of materials that have unique physical, mechanical and functional properties. They are comprised of at least two phases: a nanocrystalline phase and an amorphous phase, or two different nanocrystalline phases. The general characteristics of nanocomposite coatings are a host material with another material homogenously embedded in it, with one (or both) of these materials having a characteristic length scale of 1–100 nm. Figure 4.1 shows an example of a nanocomposite obtained from a powder of Fe–1.5 wt.% Mo (FeMo), produced by water atomization, which has been high-energy milled to obtain a nanostructure, then mixed with 1.5 wt.% of nanometric SiO_2 powder with a mean particle size of 10 nm and purity better than 99.5 per cent; and finally the mixture has been consolidated by Spark Plasma Sintering (SPS) under different conditions. Oxide particles are added to control grain growth (Cabibbo, 2009).

The behavior of nanocomposite materials mainly depends on the bulk of the grains in which dislocations play a decisive

Figure 4.1 Example: Fe-Mo nanostructured alloy

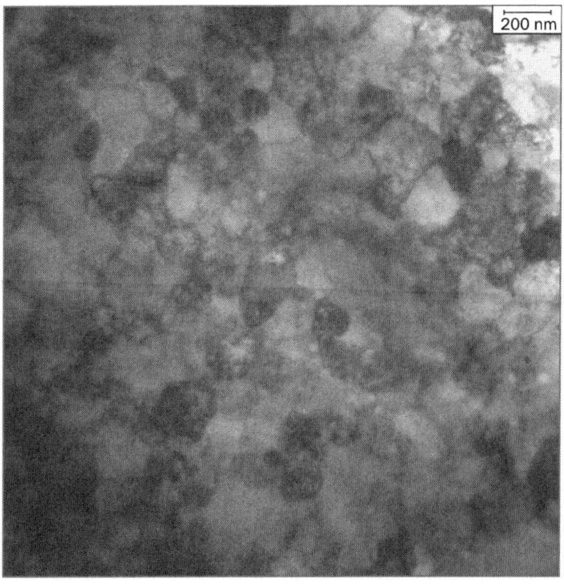

role. Completely new physical, functional and service properties are exhibited by nanocrystalline materials with a grain size of about 10 nm or less (Musil, 2000). In this case, when the size *d* of grains decreases to approximately 10 nm or less, the number of atoms in the boundary region, which surrounds the grain, is comparable to or even greater than that in the grain. This strongly changes the interaction between grains, hinders the generation of the dislocations and prevents the propagation of cracks due to the grain boundary enhancement and suppression of grain boundary sliding. The macrostress generated in the coating strongly decreases.

Nanocomposite coatings therefore possess unique mechanical and tribological properties, exhibiting hardness significantly exceeding that given by the rule of mixtures, i.e.

$$H(AaBb) = \frac{aH(A) + bH(B)}{a+b} \qquad [4.1]$$

where $H(A)$ is the hardness of pure A, $H(B)$ the hardness of pure B, a is the composition of A in the mixture and b is the composition of B in the mixture. $H(AaBb)$ is the hardness of the mixture, and the hardness of a material is usually defined as:

$$H = \frac{P_{max}}{A} \qquad [4.2]$$

where P_{max} is the maximum applied load, and A is the surface or projected area of the impression.

The described behavior mainly depends on grain boundary effects, and on the synergetic interaction of the composite constituents. Other mechanical properties, such as high elastic modulus and recovery, excellent resistance against cracking, low wear rate and high thermal stability, are due to their unique structures and deformation mechanisms at the nanometer scale (Zhang et al., 2007). Hardness is surely the most important mechanical property of nanocomposite coatings, and it is probably for this reason that they are generally classified as 'hard', 'superhard' or 'ultrahard' for hardness over 20, 40 or 100 GPa, respectively.

Examples of nanocomposites are nc-MeN/hard matrix (e.g., a-Si_3N_4, a-TiB_2, etc.) and nc-Me/soft matrix (e.g., Cu, Ni, etc.), where nc- and a- denote the nanocrystalline and amorphous phases, respectively, and Me = Cr, Ti, W, V, Zr, etc., are transition metals. The hardness H of the coatings of such materials can vary from low values of about 10 GPa to very high values up to 50–100 GPa. However, for engineering applications, hardness must be complemented with high toughness.

In the evaluation of the performance of different coatings, it must be borne in mind there is no single 'best' coating. The performance depends on many factors, and a coating which for example performs excellently in one machining operation may fail in another. Therefore, extensive testing is always needed to obtain the most suitable coating for a given application. Precision control and determination of grain size is very important for these new materials.

Modeling and simulation to explain physical mechanisms and microstructural representation of nanocomposite coating based on analysis at different levels from continuum to atomic scales, are discussed in Chunsheng and Yiu-Wing (2006), and in Pearson et al. (2009). These are very useful tools in material design.

Nanocomposite coatings with their unique properties are therefore very attractive for many industrial applications in high-speed machining, tooling, optical applications and magnetic storage devices, and in the automotive, aerospace and biomedical sectors, among others because of their special mechanical, electronic, magnetic and optical properties due to size effect.

Nanocomposite coatings of a different nature, hybrid organic–inorganic, are also studied and applied in other industrial fields, such as construction and building (e.g., wood flooring), sports, optoelectronics and mechanics. They strongly differ in physical, functional and mechanical properties from those discussed above, but the general characteristics of nanocomposite coating are the same, including a host material with another material homogenously embedded in it, having a characteristic length scale of nanometric dimensions. Generally, in inorganic–organic materials, the organic component provides flexibility for the coatings whereas the inorganic component influences the hardness and mechanical properties. Nanosized inorganic components improve scratch and wear resistance, maintaining low viscosity, gloss and transparency of the coatings (Bautista et al., 2011). The preparation of these coatings is a difficult task because it is based on the effective and uniform dispersion of the inorganic nanofillers in the polymeric matrix avoiding aggregation and phase separation.

A brief discussion of nanocomposite polymer coatings is provided in the following chapter.

4.2 Main synthesis methods

Nanocomposite coatings can be prepared by different techniques and the specific deposition method is chosen on the basis of the coating application, thickness, desired coating properties, cost and production, temperature limitation of the substrate, etc. The most promising methods are magnetron sputtering and chemical vapor deposition (CVD), although other methods, such as thermal evaporation, ion beam deposition, ion implantation and laser ablation, are also used (Zhang et al., 2003). For high deposition rate and uniform deposition in the case of complicated geometries, the CVD method is advantageous compared to sputtering.

In the CVD method, the precursor gases, $TiCl_4$, $SiCl_4$ or SiH_4, could pose problems in production because of their corrosive nature and potential fire hazard. Moreover, the chloride incorporated in protective films deposited by plasma CVD can induce interface corrosion problems during exposure to high temperatures under working conditions. In many applications, a low deposition is required to prevent substrate distortion and loss of mechanical properties. This is difficult to realize in CVD processes. Magnetron sputtering and pulsed-laser deposition can be combined to fabricate superhard and tough nanocomposite coatings (Zhang et al., 2003).

Reactive magnetron sputtering is a low-temperature technology, a far less dangerous method compared with CVD, and it can easily scale up for industrial application. This method is used to synthesize a wide range of hard or superhard nanocomposite coatings, depositing films with controlled texture and crystallite size. In this process many process parameters affect the grain size of the coating including temperature and ion current density of substrate, bias voltage, partial pressure of reactive gas and post-annealing temperature. The target power density and bias voltage have to be optimized, low-energy ion bombardment is preferred, the kinetic energy of bombarding atoms/ions is transferred into small areas of atomic dimensions and quickly converted into their close vicinity, generating extremely fast cooling and producing dense films. The crystallite size generally increases with the increase in substrate bias. In the transition metal-nitride coatings, such as Ni-Cr-N and Zr-Y-N systems, crystallite size in the coating strongly depends on partial pressure of nitrogen. Moreover, the concentration of the alloy element (Cu concentration in

TiN-Cu, Si concentration in nc-TiN/a-Si_3N_4, Si_3N_4 fraction in nc-TiB/a-Si_3N_4 and nc-W_2N/Si_3N_4) has a relevant effect on the size of the crystallites.

Magnetron sputtering is so used to obtain deposit films with controlled texture and crystallite size. The growth of crystalline phase to the required diameter and the sufficient diffusion within the segregating system can be promoted by using a minimum temperature.

For large scale industrial applications, the physical vapor deposition (PVD) methods based on sputtering and cathodic arc evaporation (CAE) are more convenient (Musil, 2001). The nanocomposite coatings are sputtered from an alloy target and it is very difficult to change continuously the chemical composition of the coatings. This can be overcome when the two-phase nanocomposite film is sputtered simultaneously from two targets of different pure elements. The regulation of the power to individual magnetrons by variation of the length and the amplitude of pulses permits the control of the chemical composition of the two-phase nanocomposite coatings. Moreover, the bipolar operation of the dual magnetron allows the deposition of oxide nanocomposite films with a high deposition rate, similar to that of pure metals. The advantages of this technology could allow the realization of new applications, e.g. in optics.

4.3 Coating properties and characterization

Nanocomposite coatings can generally exhibit high hardness, high toughness, low friction, high adhesion strength on substrate, good load support capability and chemical and thermal stability. Moreover, conventional hard materials are brittle and undergo fracture at a strain of ≤ = 0.1 per cent;

instead, in nanocomposites an unusual combination of very high hardness of 40–100 GPa, high elastic recovery of 80–94 per cent, and a high resistance against crack formation are found. This can be understood in terms of the conventional fracture physics and mechanics scaled down to a crystallite size of 3–5 nm. The superhardness is a consequence of the formation of nanocomposite with a strong interface practically without grain boundary sliding, as discussed above. The high resistance against crack formation is due to the small stress concentration factor of 2–4 in the nanocomposites, much smaller than that of ordinary microcrystalline materials. For this reason, the yield stress needed to propagate a nanocrack is much higher in nanocomposites than in a microcrystalline material, even if they have the same stress intensity factor. The high elastic recovery up to 94 per cent and high range of predominantly elastic deformation of ≥ = 10 per cent, which results in a very high elastic energy density upon indentation, can be understood in terms of reversible flexing (Veprek and Jilek, 2002).

Hardness is surely the most important property, especially in tribological industrial applications. The high hardness in nanocomposite originates from their nanostructure and remains stable to high temperatures upon annealing. In coatings such as CrN/Ni, ZrN/Ni and others, the hardness is due to a high compressive stress induced in the coatings, which is generally dependent on the energetic ion bombardment during their deposition (Veprek and Jilek, 2002). Figure 4.2 shows the hardness of a material as a function of the grain size: with a decrease in grain size the multiplication and mobility of dislocations are hindered, and the hardness of materials increases according to the well known 'Hall–Petch' relationship.

In order to evaluate hardness for thin films and coatings, nanoindentation is the best method. In this test a diamond

Figure 4.2 Hardness of a material as a function of the grain size

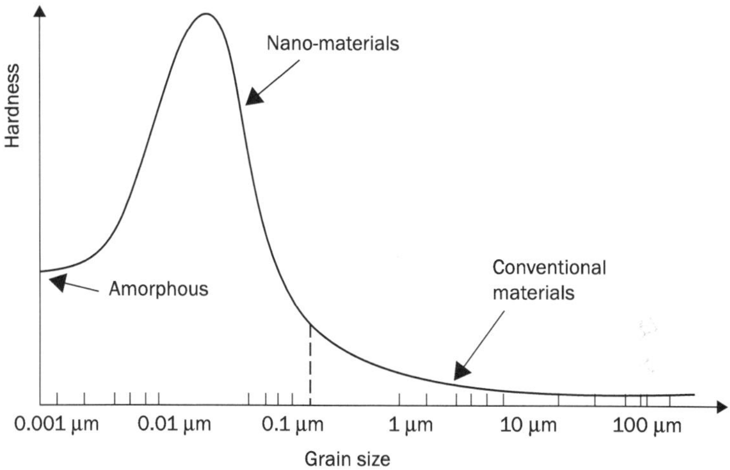

indenter is forced into the coating surface to record the load as a function of the depth of penetration. The harder a material, the more difficult it is to make an indentation or scratch. Traditional methods of measuring hardness, such as Vickers, Brinell, Knoop, Berkovich and Rockwell tests, although simple and convenient, involve a lot of variables to provide a scientific definition of hardness. Hence, an absolute scale for hardness is not available, and the different types of tests measure different things (Chunsheng Lu and Yiu-Wing Mai, 2006). For example, in the case of the Vickers test, A is the total surface area of indentation (see Eq. [4.2]), and the Vickers hardness is approximately equal to the average pressure under an indenter at yielding.

The instrumented nanoindentation test has been widely applied, not only for hardness measurement but in testing mechanical properties such as elastic modulus and fracture toughness of thin films and small volumes of material, in the three fundamental modes of deformation in solids – elasticity, plasticity and fracture. The mechanical properties can be directly determined from indentation load–displacement curves without the need to image the indent impression.

Different methods of hardness calculation can be used: the Oliver–Pharr method, the deformation energy method, the force indentation function method, the Joslin–Oliver method and the energy density method (Zhang et al., 2007). The hardness value of the same coatings thus depends on the evaluation method that must be presented when presenting hardness data. The hardness of a coating is also affected by residual stress generated during deposition, usually a few GPa for superhard coatings. This could cause significant increase in the hardness value, even if for some coatings an annealing reduces the effect.

Using the nanoindentation method introduced by Oliver and Pharr (1992) to measure the hardness and elastic modulus of a nanocomposite coating, three quantities must be determined from the load–displacement curves during one cycle of loading and unloading: the maximum load P_{max}, the maximum displacement h_{max}, and the elastic unloading stiffness $S = dP/dh$. Thus, the determination of hardness and elastic modulus are mainly based on the unloading curve (Chunsheng Lu and Yiu-Wing Mai, 2006).

Even if this method has facilitated the measurement of mechanical properties at small scales, in this case many factors influence the results. External factors include indenter geometry, tip rounding, test machine compliance, etc., and factors depending on material such as pile-up and sink-in, indentation size effect, surface roughness, residual stress, thickness of the coating, substrate, etc. This has led to claims that superhard or even ultrahard ($H > 70$ GPa) coatings are caused by artifacts in synthesis or measurement.

For quantitative evaluation of fracture toughness, there is not yet a standard method (Zhang et al., 2007), and an ultra-low load indentation is currently used. If no cracking occurs after the indentation, the coating is said to have good toughness. In any case, a quantitative description requires measurement of crack length, which is very difficult in thin films. In this method, the nature of the indenter is very important. If the indenter is not sharp enough (for example the Vickers indenter widely used), cracking does not occur even under relatively high load. In this case, the coating is defined as 'tough', but improperly so, because the resulting indentation stress does not reach the cracking threshold. This occurrence was found in most brittle materials. There is another method to evaluate toughness, which is based on energy release in through-thickness cracking in a coating obtained from a step observed in the indentation profile. Such a method was used to evaluate the fracture toughness of amorphous thin films.

Nanoindentation with a Berkovich indenter has been used to measure the coating hardness (H) and the effective Young's modulus (E), and a cube corner indenter employed to probe the coating toughness, as Galvan et al. (2006) have shown for nanocomposite coatings based on TiC nanoparticles embedded in an amorphous hydrocarbon (a-C:H) matrix, deposited via reactive closed field unbalanced magnetron sputtering.

The method to assess the adhesion of coating to the substrate is instead the scratch adhesion test. This test only reveals the load bearing capacity of the coating, rather than the real adhesion between the coating and the substrate. Many factors influence the results: substrate hardness, coating surface roughness, coating hardness, coating thickness, loading rate, indenter geometry, in particular indenter radius, and the friction between coating and indenter. There are two ways to improve coating adhesion: the first is to add a functionally graded layer; the second is to bond a layer in between the coating and the substrate.

A good interlayer bonding can be added; this provides the relaxation and modification of the stress distribution, a support layer, better chemical stability in the substrate and an increase in hardening depth (Zhang et al., 2007).

In order to assess the tribological behavior of nanocomposite coatings, a tribometer can be used. This determines the magnitude of friction and wear as two surfaces rub together. Such tests can be also performed in different environments (high vacuum, dry nitrogen and humid air), particularly, for nanocomposite coatings for aerospace applications (Wu et al., 2004; Voevodin et al., 1999).

To better understand and improve the properties of nanocomposite coatings, characterization of the chemical composition, phase composition and nanostructure are fundamental. The chemical composition can be determined using different techniques such as X-ray photoelectron spectroscopy (XPS), Auger electron spectroscopy (AES), Rutherford backscattering spectrometry (RBS), energy dispersive X-ray analysis (EDX), etc. The nanocrystalline phase can be identified by using diffraction methods such as X-ray diffraction (XRD) or electron diffraction. Average grain size can be determined from analyzing XRD peaks and applying the formula of Scherrer or the Warren–Averbach approach. The nanocrystallite size and spatial distribution within the coating bulk can be examined by using transmission electron microscopy (TEM). When the grain size is less than about 2 nm, the XRD peaks become strongly broadened, resulting in the spectrum often being difficult to interpret and the coating nanostructure is often termed 'X-ray amorphous'.

Diffraction techniques offer little or no information on the presence or composition of phases in the coatings that are either X-ray amorphous or truly amorphous. Other techniques can be used to determinate the presence and composition of such phases, the most useful being XPS, but some information can also be obtained using optical methods such as Fourier transform infra-red spectroscopy (FTIR) or Raman spectroscopy. Diffraction techniques are also unable to provide information on the relative phase fraction, because diffraction spectra are inherently non-quantifiable. XPS can provide a good method for determining relative (atomic) phase fractions, because both nanocrystalline and amorphous phases can often be separately identified and the spectra easily quantified. Baker (2007) has described how microscopic, diffraction and electron spectroscopic techniques can be employed to give a reasonably comprehensive characterization of the structure and relative phase fraction of nanocomposite coatings, analyzing in particular the Ti-Al-B-N, Ti-B-C and Cr-Cu N coating systems.

4.4 Nanocomposite polymer coatings

Nanocomposite polymers are a class of polymers reinforced with a small amount (typically 1 to 10 per cent in weight) of particles with nanometric size of different chemical nature (organic, inorganic, hybrid). Recently, they have attracted increasing attention because of their unique physical and chemical properties generally resulting from the combination of organic and inorganic materials in a single compound. They possess enhanced optical, mechanical, magnetic and optoelectronic properties. Such properties can be tuned by varying the composition, size and concentration of inorganic particles. They can be prepared in bulk, coatings or films using different methods. Examples of nanocomposite polymer coatings are nano-titania/epoxy, nano-clay/polyester, antimony-doped tin oxide (ATO) nanoparticles/poly(vinyl acetate-co-acrylic acid), Zn nanoparticles/polyaniline, $BaTiO_3$/polydiacrylate, nano-SiO_2/acrylic resin, nanosized silica/(meth) acrylate, and polytetrafluoroethylene/Cu.

The preparation of these coatings is a difficult task because it is based on the effective and uniform dispersion of the inorganic nanofillers in the polymeric matrix avoiding aggregation and phase separation, as mentioned above.

4.5 Recent developments and industrial applications

In recent years many studies have been carried out on the developments of nanocomposite coatings and films, and on their possible industrial applications in different areas such as automotive, aerospace, electronics, biomedical, sensors and bioceramics. Some studies of interest are briefly discussed.

Superhard nanocomposite coatings are already applied on machining, stamping and forming tools. They have significantly increased the lifetime (about 85 per cent higher than the optimized TiAlN coatings when applied to high-speed machining of hardened steel) and the cutting speed, and consequently also the overall productivity of the machining operations. Most widely investigated are nc-TiN/a-S_3N_4 nanocomposites (nc- stands for nanocrystalline, a- for X-ray amorphous) (Veprek and Veprek-Heijman, 2008). A further important advantage of nc-TiN/a-S_3N_4 nanocomposites is their high oxidation resistance up to temperatures of more than 800°C. Settineri and Faga (2006) have characterized two of these innovative nanostructured coatings. Scratch test, adhesion and tribological tests were carried out, and experimental results have given an important indication of their functional quality.

An interesting application of nanocomposite coatings is for hard and corrosion resistant coating of Al alloy. In the coating of aluminum alloy, micro-arc discharge oxidation, glass-flame spray, plasma thermal spray, physical vapor deposition, electroplating and high-temperature glass enameling methods are traditionally used. Recently, Abdel (2008) has produced nanocomposite coatings by electrochemical deposition of the matrix material from a solution containing suspended particles of TiO_2. In particular, an Ni-TiO_2 nanocomposite layer is deposited on Al alloy substrate by conventional direct current electrodeposition process. Ni-TiO_2 nanocomposite coatings exhibit higher corrosion resistance in comparison with a pure Ni coating. SEM observations show that TiO_2 nanoparticles are uniformly distributed into the Ni matrix. The Ni-TiO_2 morphology is smoother, more uniform and compact in appearance than Ni coating. Significant improvements in microhardness and erosion wear properties of the coating are found.

In aerospace, extreme mechanical, environmental and endurance requirements of systems require advanced friction and wear protective coatings with unique tribological performance. The WC/DLC/WS_2 nanocomposite coatings, named WCS, have demonstrated a good combination of hardness, wear resistance and low friction in humid air, dry nitrogen and vacuum. WCS consists of nanoparticles of hard WC and solid lubricant WS_2 embedded in a non-hydrogenated diamond-like carbon DLC matrix. In particular, WCS nanocomposite coatings are designed to have excellent tribological properties (low friction and wear, long life and the ability to adapt to changes in ambient conditions during sliding) (Wu et al., 2004).

Recently, the great interest in preserving the cultural heritage (architectural and sculptural stone), has led to the development of new protective materials. Coating in stone conservation has mainly three aims: to protect undamaged stone, to consolidate and preserve damaged stone and to slow down their ageing process. The best chemical products are silicon derivatives that are widely used in stone conservation for their low viscosity and deep penetration within the stone, as well for their good weather resistance and water vapor permeability characteristics. A new nanocomposite coating, silicon-based hybrid polymer nanocomposite with silsesquioxane units (TMSPMA), has been proposed as a protective material for two porous limestones, i.e., a bioelastic limestone with large pore radii and grain size, and a dolomitic limestone of small pore radii and grain size, both used as monumental building stone (Simionescu, 2009). The new coating acts as a water repellent or as consolidant, function of stone composition, structure and porosity. Upon UV ageing at 40°C, TMSPMA polymer layers become more hydrophobic, also ensuring that the stone will exhibit resistance and durability over time against UV radiation.

Nanocomposite polymer coatings are also being increasingly investigated for the production of multifunctional textiles. A novel technique for producing durable multifunctional textile using nanocomposite coatings has been proposed (Dastjerdi et al., 2010). Different concentrations of TiO_2 and Ag nanosized colloidal solution and polysiloxane resin as stabilization compound for nanoparticles on textile surfaces (polyester fabric) are used. The choice of polyester is due to its special advantages and a wide range of applications of polyester in the textile industry, especially for medical and sport uses. Characteristics of different treated fabrics including hydrophilicity–hydrophobicity, antibacterial efficiency, anti-staining properties, stain photo-degradability, UV protection, air permeability and washing durability are evaluated in this study. Ideal antibacterial, self-cleaning, anti-staining properties can be achieved by applying a mixture of 300 ppm silver and 1 wt% cross-linkable polysiloxane (XPs) as an after-treatment on 2 wt% TiO_2, as well as 5 wt% TiO_2 with the same after-treatment. The desirable hydrophilicity can be reached by controlling the chemical ratios. The major stabilizing effect of XPs and its ability to counteract some adverse properties of inorganic particles, such as decreasing conductivity and softness, and also to offer protection to the human body from direct contact with nanoparticles, is intensified by the photo-catalytic activities of TiO_2. Finally, an optimum UV protection is recorded.

Recently, studies on polyester–clay nanocomposites have been performed for the interest of this class of materials, mainly in medical and sanitary applications, and packaging (for material transparency and oxygen barrier properties). One of these studies proposes a new manufacturing procedure for industrial application, strongly reducing the process time (less than 1 h to obtain thick films of 250 μm), (Lucignano et al., 2008). Nanocomposite thick films are prepared by the in situ intercalative polymerization method using nano-clay powder and unsaturated polyester (PU) matrix. The nanofiller used is montmorillonite (MMT). The nano-clay derived from especially purified and natural MMT was modified with quaternary ammonium salt. Particle size of nano-clay was expected to be about 7–9 nm.

Figure 4.3 shows a scanning electron microscopy (SEM) image of the nano-powder. Experimental results show that a maximum nanocomposite mechanical performance was observed at 1 wt% filler content. This is very attractive considering the fast manufacturing procedure. Both the tensile storage modulus and the tensile strength show an increase of about 150 per cent in comparison with the unfilled polyester resin. Figure 4.4 shows the tensile strength of the samples as a function of clay content.

Figure 4.3 SEM image of nano-powder

Figure 4.4 Tensile strength against nano-clay content

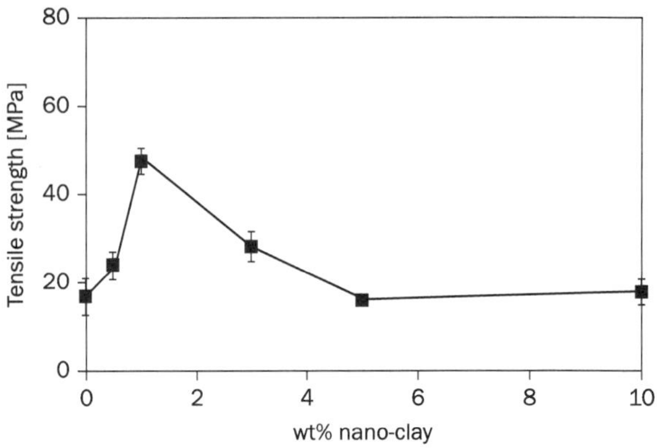

4.6 Future trends

Different points of view are expressed in literature on future trends. Pogrebnjak et al. (2009) have pointed out that further research in the field of nanostructured films and coatings will be performed mainly along the following lines:

1. Coating or film formation with controllable grain size in the range of 1 to 10 nm, for the explanation of the

relationship between their properties and dimensional parameters, and the production of new improved coatings with unique physical and functional characteristics.

2. Nanocrystallization from the amorphous phase
3. The transfer of electron charges between nanograins of different chemical composition and Fermi energy.
4. The preparation of new protective coatings with an oxidation resistance higher than 2000°C.
5. The development of new deposition systems of nanostructurate coatings by physical methods.

Some materials like TiN, TiNCr and TiZrN coatings and TiN/CrN, MoN/TiAlN multilayer coatings, and others, are being used in various industrial applications. The addition of other components (e.g., Si or B), into TiN films can significantly improve their physical and mechanical properties, increasing possible applications. It is shown that by doping classical TiN coatings with Si, B, Al or Cr, a combination of high hardness and durability with a sufficient low coefficient of friction is obtained. These multi-component nanostructured coating and films can be used for protection of different components and instruments, such as cutting and stamping tools, casting rollers, parts of aircraft engines, gas-driven turbines and compressors, sliding bearings and extrusion nozzles for glass and mineral fibers. They can act against the simultaneous actions of high temperatures, mechanical wear and aggressive media. Other promising fields of application are magnetic optics, electronic engineering and medicine. In medicine, coatings are necessary to develop a new generation of biocompatible materials, e.g., orthopedic and dental implants, materials for cranial and maxillofacial surgery and repair of the cervical and lumbar spine (Pogrebnjak, 2009).

Zhang et al. (2007) observed that the main future target will be to obtain nanocomposite coatings with a combination of hardness and toughness. This is driven by industrial demand. In principle, all toughening mechanisms can be applied to strengthening ceramic thin films or coatings in particular; the real challenge lies in implementation. In situ synthesis of carbon nanotube (CNTs)-imbedded nanocomposites, such as nc-TiC/a-C(CNT) or nc-TiN/a-Si_3N_4(CNT) or nanocrystals imbedded in a complete matrix of carbon nanotubes (for instance nc-TiC, or even nc-Diamond/CNT), may hold the key to this ideal 'hard yet tough' coating. In this case the CNT-imbedded matrix, or the CNT matrix, provides the ultimate 'bulk' elasticity and toughness to stop crack propagations, and the nanocrystalline phase provides the higher hardness. One limiting factor in the progress of this research is surely the lack of a universally accepted toughness measurement methodology and apparatus for thin films and coatings. It is hoped that soon a universally accepted methodology and a standard test procedure will be found.

4.7 References

Abdel Aal A (2008), 'Hard and corrosion resistant nanocomposite coating for Al alloy', *Material Science and Engineering A*, **474**: 181–7.

Ali Olad, Haleh Rasouli (2010), 'Enhanced corrosion protective coating based on conducting polyaniline/zinc nanocomposite', *Journal of Applied Polymer Science*, **115**: 2221–7.

Baker M A (2007), 'Advanced characterisation of nanocomposite coatings', *Surface & Coating Technology*, **201**: 6105–11.

Bautista Y, Gonzales J, Gilabert J, Ibanez M J, Sanz V (2011), 'Correlation between the wear resistance, and the scratch resistance, for nanocomposite coatings', *Progress in Organic Coatings*, **70**: 178–85.

Cabibbo M, Deodati P, Libardi S, Molinari A, Montanari R, Ucciardello N (2009), 'Damping of FeMo alloys obtained from SPS sintering of nanostructured powders', *Materials Science Forum*, **604–5**: 203–11.

Chen K Y, Wang C A, Huang Y, Lin W (2009), 'Preparation and characterization of polymer–clay nanocomposite films', *Science in China Series B: Chemistry*, **52**: 12, 2323–8.

Chunsheng Lu, Yiu-Wing Mai (2006), 'Recent advances on the understanding the origin of superhardness in nanocomposite coatings: A critical review', *J. Mater Sci*, **41**: 937–50.

Dastjerdi R, Montazer M, Shahsavan S (2010), 'A novel technique for producing durable multifunctional textiles using nanocomposite coating', *Colloids and Surfaces B: Biointerfaces*, **81**: 32–41.

Galvan D, Pei Y T, De Hosson J Th M (2006), 'Influence of deposition parameters on the structure and mechanical properties of nanocomposite coatings', *Surface & Coating Technology*, **201**: 590–8.

Holubar P, Jilek M, Sima M (2000), 'Present and possible future applications of superhard nanocomposite coatings', *Surface and Coating Technology*, **133–4**: 145–51.

Li H, Chen Y, Ruan C, Gao W, Xie Y (2001), 'Preparation of organic–inorganic multifunctional nanocomposite coating *via* sol gel routes', *Journal of Nanoparticle Research*, **3**: 157–60.

Lik Hang Chau J, Tung C-T, Lin Y-M, Li A-K (2008), 'Preparation and optical properties of titania/epoxy nanocomposite coatings', *Materials Letters*, **62**: 3416–18.

Lucignano C, Quadrini F, Santo L (2008), 'Dynamic mechanical performances of polyester–clay nanocomposite thick films', *Journal of Composite Materials*, **42(26)**: 2841–52.

Musil J (2000), 'Hard and superhard nanocomposite coatings', *Surface and Coating Technology*, **125**: 322–30.

Musil J, Polakova H (2000), 'Hard nanocomposite Zr-Y-N coatings, correlation between hardness and structure', *Surface and Coating Technology*, **127**: 99–106.

Musil J, Vlcek J (2001), 'Magnetron sputtering of hard nanocomposite coating and their properties', *Surface and Coating Technology*, **142–4**: 557–66.

Oliver W, Pharr G (1992), 'An improved technique for determining hardness and elastic modulus using load and displacement sensing indentation experiments', *Journal of Material Research*, **7**: 1564–83.

Pearson J D, Zirky M A, Wahl K J (2009), 'Microstructural modeling of adaptive nanocomposite coatings for durability and wear', *Wear*, **266**: 1003–12.

Pogrebnjak A D, Shpak A P, Azarenkov N A, Beresnev V M (2009), 'Structure and properties of hard and superhard nanocomposite coatings', *Physics Uspekhi*, **52**(1): 29–54.

Rogachev A, Yarmolenko M, Rahachou A V, Tamulevičius S, Prosycevas I (2008), 'Structure and properties of nanocomposite polymer coatings', *Journal of Physic: Conference Series 100*, 082042.

Schmittegens R, Wolf M, Schultheiss E (2009), 'A versatile system for large area coating of nanocomposite thin films', *Plasma Processes and Polymers*, **6**: 5912–16.

Settineri L, Faga M G (2006), 'Laboratory tests for performance evaluation of nanocomposite coatings for cutting tools', *Wear*, **260**: 326–32.

Sheinman E (2008), 'Coating from nanocomposites', *Metal Science and Heat Treatment*, **50**: 601–5.

Simionescu B, Aflori M, Olaru M (2009), 'Protective coatings based on silsesquioxane films for building limestones', *Construction and Building Materials*, **23**: 3426–30.

Solarski S, Benali S, Rochery M, Devaux E, Alexandre M et al. (2005), 'Synthesis of a polyurethane/clay nanocomposite used as coating: interactions between the counterions of clay and the isocynate and incidence on the nanocomposite structure', *Journal of Applied Polymer Science*, **95**: 238–44.

Soloukhin V A, Posthumus W, Brokken-Zijp L C M, Loos J, de With G (2002), 'Mechanical properties of silica-(meth) acrylate hybrid coatings on polycarbonate substrate', *Polymer*, **43**: 6169–81.

Sun Q, Schork F J, Deng Y (2007), 'Water-based polymer/clay nanocomposite suspension for improving water and moisture barrier in coating', *Composites Science and Technology*, **67**: 1823–9.

Sun J, Francis L F, Geberich W W (2005), 'Mechanical properties of polymer–ceramic nanocomposite coatings by depth-sensing indentation', *Polymer Engineering and Science*, **45**(2): 207–16.

Tran T A, Saïd S, Grohens Y (2005), 'Compared study of cooperativity in PMMA nanocomposites and thin films', *Composites: Part A*, **36**: 461–5.

Tsaneva G, Kokhuzharov V, Kokhuzharov S, Ivanova M, Gerwann J et al. (2008), 'Functional nanocomposite coatings for corrosion protection of aluminium alloy and steel', *Journal of the University of Chemical Technology and Metallurgy*, **43**(2): 231–8.

Veprek S, Veprek-Heijman M J G (2008), 'Industrial applications of superhard nanocomposite coatings', *Surface & Coating Technology*, **202**: 5063–73.

Veprek S, Jilek M (2002), 'Superhard nanocomposite coatings. From basic science toward industrialization', *Pure Appl. Chem*, **3**: 475–81.

Voevodin A A, Zabinski J S, Muratore C (2005), 'Recent advances in hard, tough, and low friction nanocomposite coatings', *Tsinghua Science and Technology*, **10**(6): 665–79.

Voevodin A A, O'Neill J P, Zabinski J S (1999), 'Nanocomposite tribological coatings for aerospace applications', *Surface & Coating Technology*, **116–19**: 36–45.

Wang Y, Lim S, Luo J L, Xu Z H (2006), 'Tribological and corrosion behaviors of Al_2O_3/polymer nanocomposite coatings', *Wear*, **260**: 976–83.

Wang Z, Han E, Ke W (2006), 'Effect of acrylic polymer and nanocomposite with nano-SiO_2 on thermal degradation and fire resistance of APP-DPER-MEL coating', *Polymer Degradation and Stability*, **91**: 1937–47.

Wu J H, Rigney D A, Falk M L, Sanders J H, Voevodin A A, Zabinski J S (2004), 'Tribological behaviour of WC/DLC/WS2 nanocomposite coatings', *Surface & Coating Technology*, **188–9**: 605–11.

Zao X, Hinchliffe C, Johnston C, Dobson P J, Grant P S (2008), 'Spray deposition of polymer nanocomposite films for dielectric applications', *Materials Science and Engineering B*, **151**: 140–5.

Zhang S, Wang H L, Ong S E, Sun D, Bui X L (2007), 'Hard yet tough nanocomposite coatings – present status and future trends', *Plasma Processes and Polymers*, **4**: 219–28.

Zhang S, Sun D, Fu Y, Du H (2003), 'Recent advances of superhard nanocomposite coatings: a review', *Surface & Coating Technology*, **167**: 113–19.

5

Thermal sprayed WC-Co coatings for tribological application

R.M. Mohanty, Non-Ferrous Materials Technology Development Centre, India, and M. Roy, Defence Metallurgical Research Laboratory, India

Abstract: The WC-Co coating obtained by thermal spraying technique is one of the most important and widely used coatings for tribological applications. In this chapter, various methods of spraying WC-Co coatings are discussed along with their microstructural features and mechanical properties. Further, the tribological performances of these coatings are described in detail. This chapter is also concerned with various tribological applications and the direction of future research.

Key words: thermal spraying, tribology, wear, erosion, abrasion, WC-Co coating.

5.1 Introduction

Tribological degradations are observed when there is a relative motion between two surfaces. Such degradation becomes severe in the presence of hostile environments or foreign bodies. The challenge in solving such degradation lies in determining the type of degradation to which the component is subjected and ways of removing material from the surface. Monitoring and diagnosing such problems enhances the service performance of systems. Surface modification technologies and their selection depend on surfaces with respect to the mechanism and processes that caused the damage in the tribosystem. The ultimate objective of tribological modification is to achieve a functional system with better tribological properties.

From a historical perspective, it can be observed that major technologies are influenced by materials to be used. The development of suitable materials that withstand given engineering environments for an indefinite period is practically impossible. In such a case, a suitable surface modification will either restore or improve the performance of a component. In many cases, the need for a coating is inevitable, as here 'a coating and a material' are tailored to resist the particular functional environment. Coatings can vary from a few to several hundred microns and can be obtained by different methods (Mohanty, 2009). Depending on the ultimate use of the component and the desired environment it has to resist, the coating, its thickness and means of deposition are decided (Mohanty, 2009). Performance of movable devices, such as those found in an aircraft system, depends on availability of advanced wear and friction reduction technologies (Chun, 2010; Ichiro and Shigeyuki, 2007; Ihn-Sung, 2001; Koji, 2000; Mainsah, 2001; McCool, 1986; Michael, 2003; Simon, 2008; Nakada, 1995; Tabor, 1951; Williams, 2000). Development of such technologies requires improved materials, advanced processes and characterization and performance evaluation technologies (Barrella and Priesta,

2003; Bin, 2010; Ram, 1996; Smart and Bockris, 1987; Stewart, 1998; Stewart, 2000; Voevodin, 1999).

Typical tribological coatings, such as WC-Co cermet coatings, are generally exposed to a wide range of contact stresses, high pressures and sliding velocities. The desirable service requirements are low friction, stable wear rate, sufficiently good adhesion and high temperature stability. The expected friction coefficient is below 0.06 to 0.4 depending on the type of friction environments, whether lubricated or dry. The typical operational parameters are sliding speeds up to 18,000 rpm, contact pressures above 1,500 MPa and temperatures from very low to 1,200 K and above.

WC-Co materials have been useful for most diverse service needs such as abrasive, sliding, fretting and erosive wear resistance of engineering components in a variety of industrial applications. In these coatings, marginal lubrication with optimal porosity acting as a lubricant reservoir is sometimes used. The composite nature of a coating, including the presence of amorphous phases, helps the surface to resist wear compared to wrought or cast materials. The friction coefficient of WC-Co is derived as a sum of contributions from metal wear and the ceramic wear mechanisms. The hard particle WC forms the wear-resistant constituent of WC-Co materials, while the cobalt provides necessary toughness, adhesion and cohesion. Tungsten carbide-cobalt (WC-Co) composite material is conventionally manufactured by an expensive technology, the sintering fabrication of metal matrix composites. The conventional processing methods are costly and are limited to the production of small components. Compared to these conventional methods there are several thermal spray techniques for particulate deposition that are convenient and economic.

This chapter is an overview of the performance of WC-Co coating deposited by thermal spraying technique as a means of protecting components against tribological degradation. This chapter will highlight recent developments in the understanding of thermal sprayed WC-Co coatings in recent times, particularly in the last few years.

5.2 Thermal spray techniques

Thermal spray is a process where metals, ceramic, cermet and selected polymeric materials are fed in the form of powder, wire and rods to a torch where the materials are heated or melted, and accelerated in the gas stream towards a substrate where they form a splat. The nature of the splat is governed by the thermal history of feed materials and ultimately defines the coating microstructure. The specified coating thickness is achieved by multiple passes of the torch. The major advantage of the thermal spray over other metallurgical or materials processes is that it provides a wide choice for developing a coating suitable for a particular application. In cases like cold spray processes, coating feeds can be applied to the surfaces without any significant heating. Thermal spraying often helps in maintaining the heat treatment conditions applied to an engineering part or preventing thermal distortion or providing a better surface and dimensional accuracies to a damaged surface. When required, the coating can be stripped out and the surface can be recoated.

Several thermal spray techniques are available for engineering or reengineering the surfaces of varying components. These are plasma spraying, high velocity oxygen fuel spraying, detonation gun spraying, laser assisted spraying, flame spraying and electric arc spraying. Some of the basic properties of these processes have been summarized in Table 5.1. The heat sources used for such processes are either combustion gases or electric

Table 5.1 **Commonly observed data on various categories of thermal spraying processes**

Spray process	Materials sprayed	Process temperature (K)	Velocity (m.s)
Flame spraying	Metallic, ceramics, fusible coatings	3350	60–70
Arc spraying	Metallic coatings	7300	100–175
Plasma spraying	Metallic, ceramics, compounds and plastics	16000	150–600
Detonation gun spraying	Metallic, selected ceramics, cermets	4500	600–800
High velocity oxygen fuel spraying	Metallic, selected ceramics, cermets	3000	600–1200

arc. In general, most of the flame spray processes are adapted to use various combinations of gases and nozzles to achieve targeted operating cost and coating properties. The gases would be acetylene, propane, methyl-acetylene-propadiene, hydrogen with oxygen or air. Flame characteristics and the available flame heat are dependent on the oxygen-to-fuel gas ratio and pressure. Cold spray and flame spray/fuse methods also belong to this category of processes. In the flame spray and fuse process, the advantage is that the coating is self-fluxing in nature with oxygen or oxides. This forms low density oxides that float to the surface thereby making a stronger substrate–coating interface. These are used in cases where severe wear with high stress on the coating interfaces are observed.

The cold spray process is based on the energy stored in high-pressure compressed gas which accelerates fine powder particles at velocities of 500–1,500 m/s. The powders are also fed though a high-pressure powder feeder. The gas exits through a specially designed nozzle (laval type convergent–divergent nozzle) at very high velocity. The powder accelerated particles are often moderately heated to a certain temperature where on impact with a substrate they deform and bond to form a solid-state coating. By these processes, oxidation of spray materials and shrinkage on cooling of coating are avoided. Whatever changes are observed in the coatings compared to feed materials are due to deformation and cold working.

The electric arc spraying process is based on the controlled arc generation with an electrically shorted metal wire. The significant difference between this process and others is that it does not use any external heat source. In electric arc spray processes the molten metal on the wire tips is atomized and the stream of compressed air or other gases propels the droplets to the substrates to be coated. The substrate heating is low and the process is less expensive compared to other processes. It uses relatively ductile, electrically conductive wire of diameter less than 2 mm. These types of coatings are used in high volume and low cost application cases such as developing low cost moulds, zinc coatings, etc. The laser spray process relies on high power and power density lasers like carbon dioxide (CO_2) lasers, solid-state lasers, e.g. Nd:YAG lasers, diode lasers with power levels up to several tens of kilowatts. An absorption enhancing pretreatment is normally necessary to enhance the process.

Research and development of laser coatings, the properties of coatings and their industrial applications has been very active in the last few years. On all the laser methods, laser surface treatment is still significantly less used than, for example, laser marking, cutting or welding. The transferred arc plasma spray methods are used where high deposition rate, high thickness per pass and metallurgical bondings are necessary. It consumes less electrical power than with a non-transferred arc process. Since it heats up and melts the substrate surface, microstructural alteration of the substrate is possible. In many practical cases this is undesirable. The cost of the processing is comparatively high and the productivity is less with respect to other thermal spray processes.

As an advanced technology, a high velocity oxygen fuel spraying (HVOF) method is used for fabricating WC-Co, and now non-transferred plasma spray methods for WC-Co are also in the developmental stage (Vinayo, 1985). Generally, such coatings with good toughness, good surface fatigue resistance and low residual tensile stress are best suited to current applications. Both of the processes can be used for development and repair of wear-resisting surfaces. Plasma spraying is a high-energy process in which a high-current arc is generated within the torch and a gas is injected into the arc chamber, where it is heated and converted into high-temperature plasma. In practice, pure argon or nitrogen is used as the primary plasma gas, together with additions of 2–40 per cent of a secondary gas (hydrogen or helium). Powdered surfacing material is injected into this plasma jet and is thus heated to a molten state and accelerated onto the substrate. Plasma temperatures higher than 15,000 K and particle velocities of up to 900 m/s have been observed.

HVOF is a high-energy process. It is based on specially designed spray guns that burn oxygen and a fuel gas (hydrogen, propane or propylene). In HVOF systems, the combustion process takes place within the gun, and the gas flow rates are much higher than in conventional flame spraying. The combination of these two factors leads to supersonic flame speeds, up to approximately 2,000 m/s, with particle velocities that may reach 800 m/s. The maximum temperature attained in these processes is around 2,700–3,200 K (Table 5.1). The major disadvantage of all these thermal spray processes is that the coating deposition occurs in a line-of-sight manner. This relies upon close proximity of the coating device to the coating site. The mechanism of bonding is mostly due to mechanical interlocking of shrinking and solidifying particles. These often need rougher surfaces and post-operation treatments.

5.3 Characteristics of WC-Co coatings

5.3.1 Microstructural features of WC-Co coatings

WC-17%Co with a range of particle size from 08 μm to 64 μm has been characterized. Typical scanning electron microscope (SEM) images of commercial WC-17%Co powders are shown in Fig. 5.1. Particle size distribution of such powders is presented in Fig. 5.2. All granulated powders are of spherical shapes with the presence of comparatively large initial porosities. Figure 5.3 shows the distribution of

Figure 5.1 SEM micrographs of the powder DIAMALLOY 2005: (a) granules; (b) single granule

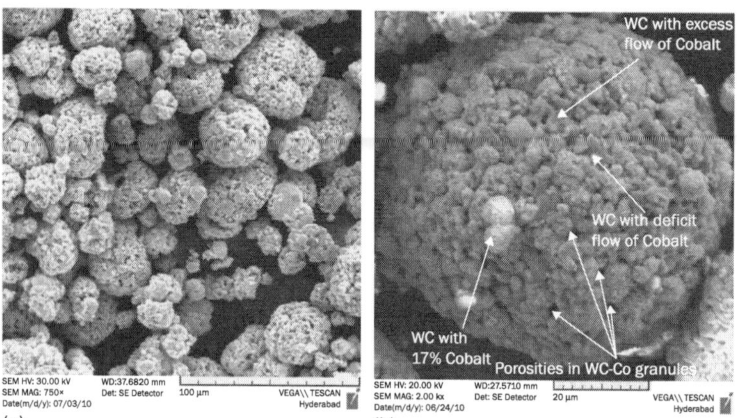

(a) (b)

Figure 5.2 Volume distribution of the powder granules: (a) measured by laser particle analyser; (b) obtained by image analysis

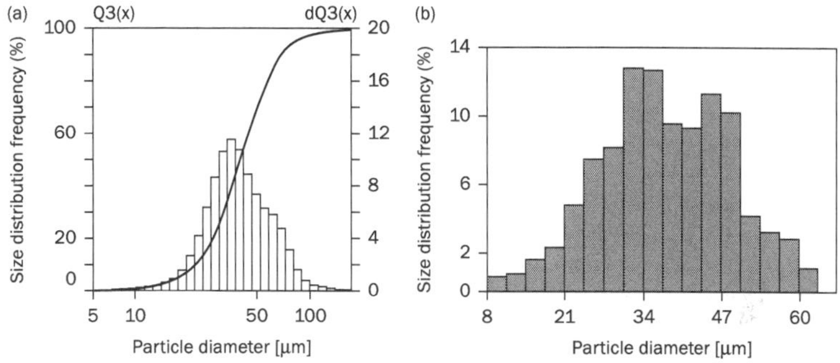

Figure 5.3 Distribution of Co, C and W in a spherical granule of DIAMALLOY 2005 with colour codes

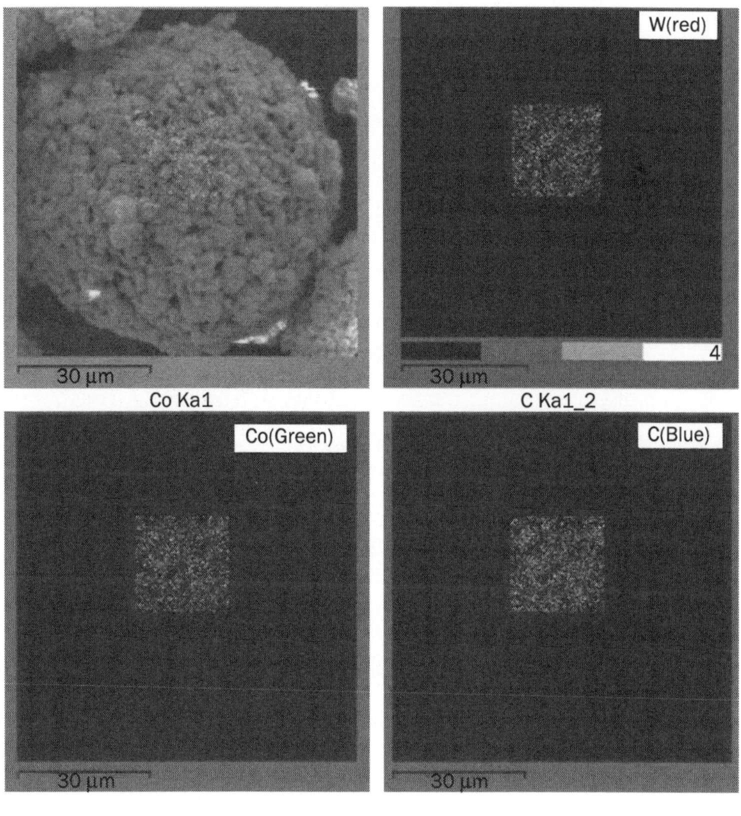

Co, C, W and O in a spherical granule of the DIAMALLOY 2005. The presence of surface oxide on the spray granules was detected by SEM analysis. There are many applications for which angular powder with different sizes and shapes is used. Angular powders are easy to produce, and if they are not sufficiently fine, the resultant coatings have porosities. Very fine powders are difficult to feed at constant rate and are sometimes responsible for a high degree of oxidation or undesirable phase transformation.

Since micro- and sub-microstructure can be of key importance, the structure and porosity analysis requires a good statistical database. The database cannot consist only of pictures of the same area; the area investigated should vary. Figure 5.4 represents the cross-sectional micro-structure of the WC-17%Co coating as observed in the plasma spray process (Fig. 5.4a) and HVOF process (Fig. 5.4b) The porosities on a WC-17%Co coated surface as determined by image analysis were observed to be less than 0.49 per cent. The related percentage of unmelted particles was less than 0.2 per cent with the exception of WC particles, which are difficult to melt. WC is one of the hardest carbides with a melting point around 3043 K. However, the hot flattening behaviour of WC is significant in the plasma spray process compared to that behaviour in the HVOF process. This phenomenon leads to the production of a dense and adherent coating by the plasma process. In such a process, the as-sprayed surface roughness (*Ra*) was about 1.5 to 2.1 μm.

Other microstructural features of coatings as produced by the well-known HVOF process for WC-Co are illustrated in Fig. 5.5 and Fig. 5.6. These figures show the results of the porosity analysis in different fields (polished surface, microscopy, etc). In porosity measurements, the type of the microscope used is important. The porosity obtained from

Figure 5.4 Micro-structure of the WC-17%Co coatings obtained by a) plasma spraying, b) HVOF spraying

(a)

(b)

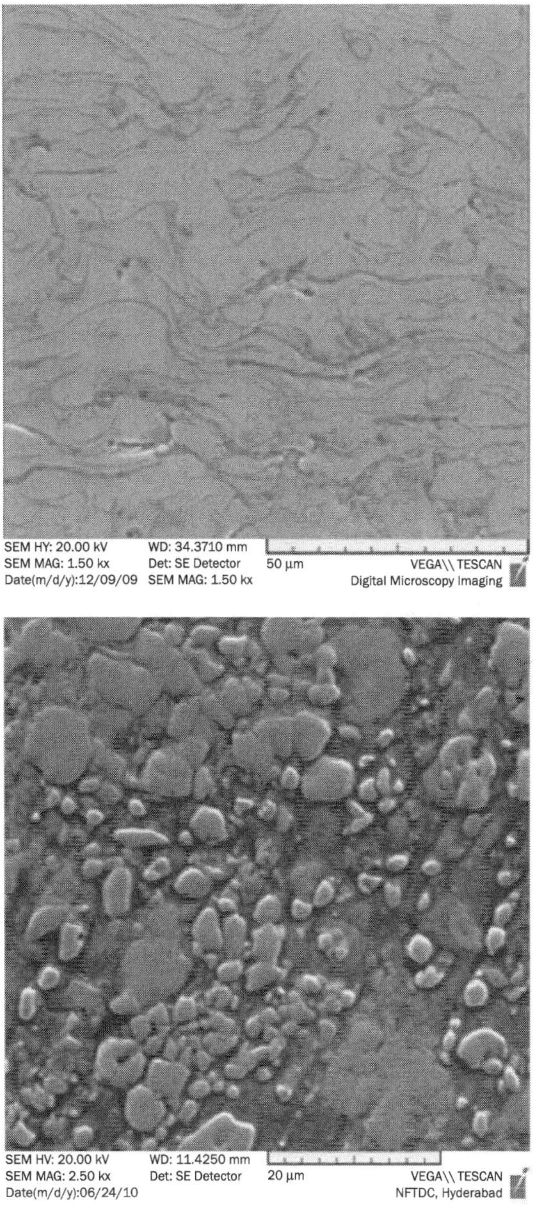

Figure 5.5 Distribution of Co, W, C and O in high magnification cross-section of the coating (70 micron × 70 micron area)

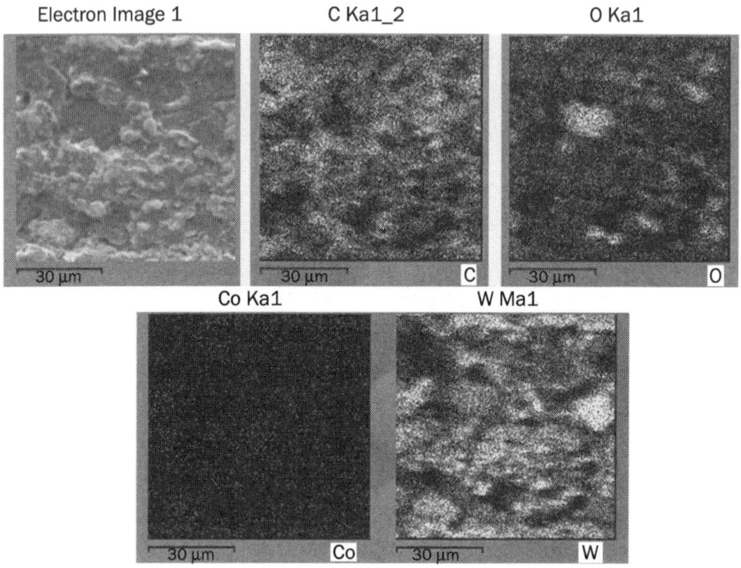

SEM pictures is much higher than that in the pictures obtained by means of an optical microscope. Microporosity is twice as high as macroporosity. It is probable that, the optical microscope did not detect microporosity while, on the other hand, SEM showed black phases that consist of lighter elements. The porosity of the WC-17%Co surface, as determined by image analysis, was observed to be well below 1.1 per cent with as-sprayed R_a of 3–4 µm, as-polished R_a of 0.1–0.5 µm and oxide content of <1 per cent. Metallographic study of the cross-section of the coating revealed a dense coating structure of less than 0.5 per cent porosity, which was transformed from the very porous starting powder agglomerates. The percentage of unmelted particle was less than 0.5 per cent. As observed in micrographs of the WC-Co

Figure 5.6 Energy-dispersive X-ray image from (70 micron × 70 micron) area of the coating

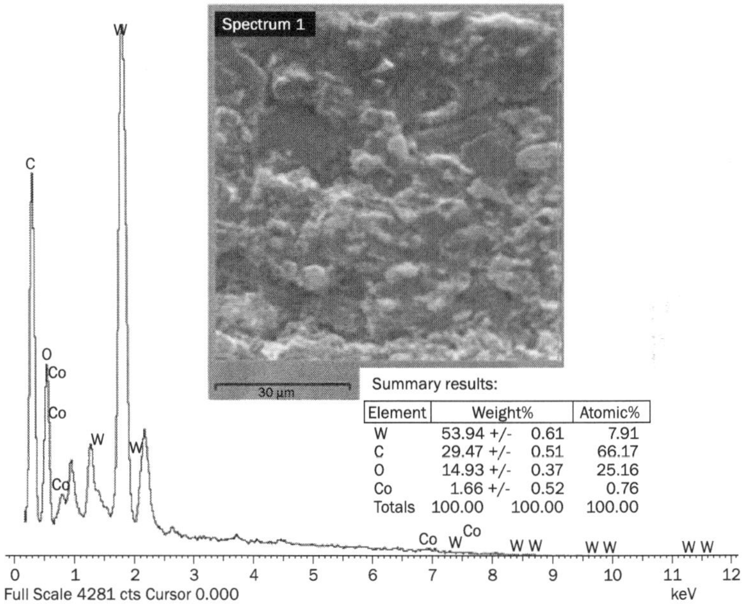

Element	Weight%	Atomic%
W	53.94 +/- 0.61	7.91
C	29.47 +/- 0.51	66.17
O	14.93 +/- 0.37	25.16
Co	1.66 +/- 0.52	0.76
Totals	100.00 100.00	100.00

by HVOF process, the primary WC has been fragmented to finer sizes.

All images from Fig. 5.4 to Fig. 5.6 reveal dense coatings with a thickness of approximately 300 microns with densely packed carbide in the binder phase. Some amount of pores were observed in all of the coatings. In some area, lack of splat–splat bonding was also observed. This indicated that in the HVOF processes the available heat transfer is still insufficient to develop very good lamellar structure as required for a thermal spray coating (Gao, 2008). It is to be stated that by HVOF process, the deposition with small powders (~10 microns) is difficult due to the spitting problem. However, the process can help in maintaining dimensional tolerances as required for severe surface restoration purposes.

5.3.2 *Hardness of the coatings*

The most important mechanical property of thermal spray coatings is hardness. Mostly, a Knoop indenter is used to measure the hardness of thermal sprayed coatings, although a Vickers indenter is used for measuring both hardness and indentation toughness. Hardness using a Knoop indenter can be given by:

$$KHN = 14229\frac{P}{a^2} \tag{5.1}$$

where *KHN* is the Knoop hardness number expressed in kg/mm^2, *P* is the applied load in g/f and *a* is the length of the major diagonal of the indenter in μm.

The microhardness of the coatings is generally evaluated across the transverse sections of the coatings at a 300 g load. Hardness value depends on the process, coating composition, etc. The hardness values of WC-12%Co coatings applied by Sulzer Metco continuous detonation gun, Sulzer Metco diamond jet, DNIPER-3 detonation gun and Miller Thermal SG-100 plasma gun are 880, 1,236, 928 and 612 (HK 0.3) respectively (Powlowski, 1995; Roy, 1999), indicating that plasma spraying results in lowest hardness. This may be related to the fact that during plasma spraying more powders undergo oxidation or undesirable phase transformation and the coating is more porous. If the powders are nanostructure, then hardness could be as high as 1570 (Sinha, 2003). Composition of the coating alters hardness, as is evident from the study of Roy et al. (2000). The hardness of detonation sprays of WC-12%Co and WC-17%Co are 928 and 754 respectively.

WC-Co coatings are normally used for tribological applications. The strain rate associated with tribological deformation is much higher than that obtained during static indentation. Thus, dynamic hardness is a better property for predicting the tribological performance of materials or coatings (Roy, 2006). The dynamic hardness of thermal sprayed WC-Co coatings is also evaluated (Roy, 2002). It is noted that dynamic hardness decreases with the increase of Co content and increase of coating thickness. Further dynamic hardness is substrate dependent. It is highest for the substrate with the lowest melting point.

Finally, the indentation toughness is an important property for coatings and can be estimated using a Vickers indenter at a given load. The indentation can be carried out on the traverse section of the coating in the mid-plane region. The indenter should be loaded so that one of the horizontal diagonals is parallel to the coating substrate interface. The diagonals of the indentation (*a*) and the cracks (*l*) formed at the corners of indentations were measured and used to estimate the characteristic crack length ($c = l + a$) and the indentation fracture toughness values (K_i). As most of the cracks are nearly Palmqvist-type cracks ($c/a < 2.5$), an expression proposed by Niihara et al. (1982, 1983) can be used. The proposed expression is:

$$K_i = 0.0123 x E^{2/5} x H^{1/10} x (P / l)^{1/2} \quad [5.2]$$

where E (N/m^2) is the Young's modulus, H (N/m^2) is the hardness and P (N) is the external load applied.

5.3.3 Bond and bend strength of the coatings

The bond and the bend strength values provide a significant in-depth evaluation of the coating along with the micro-structural, hardness and phase evaluations. Figure 5.7 represents samples used for bond testing and bend testing.

Figure 5.7 Bond test sample for testing for WC-17%Co plasma coating and bond testing as carried out for HVOF-based WC-17%Co coating

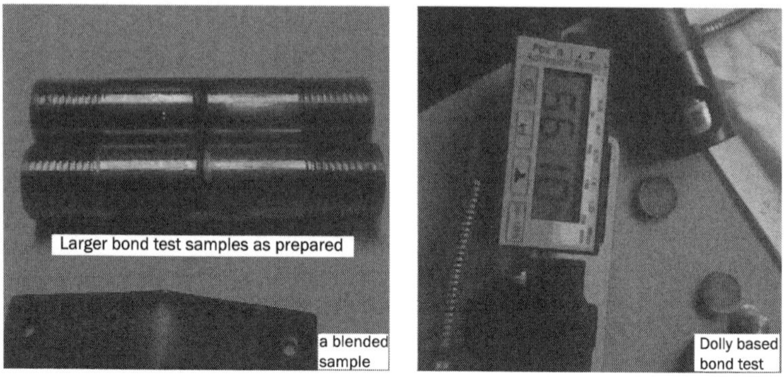

No spallation was observed during bend testing of the plasma-based WC-17%Co samples. The related bond strength was more than 36.95 MPa at the point where the adhesive delaminated. The average microhardness of the coating was around 1,187 HV0.3. The microhardness of the HVOF-based WC-17%Co coatings was 1,051 HV0.5 with an average variation of less than 2 per cent. The bond strength achieved 27–36 MPa before the adhesive delaminated. No spallation, debonding or cracking were observed during bend depths of 15–20 mm in the coating side. Bond strength in excess of 80 MPa is reported for HVOF-sprayed WC-Co coatings (Powlowski, 1995). For a series of plasma sprayed coatings, a bond strength between 588 Kg/cm^2 and 886 Kg/cm^2 is reported by Kim et al. (1994).

5.3.4 Phase transformation during spraying and post processing

WC-Co coatings undergo phase transformation during spraying and also during post-coating treatment. According to X-ray diffraction analysis, powder granules normally have the following additional phases: W-C-O and Co_3W_3C. The coatings obtained by HVOF spraying have the same phases, except that additional W and $W_2C_{0.8}$ appeared as a result of phase changes during the spraying process (Kim, 1994; Roy, 2005). A typical X-ray diffraction (XRD) pattern obtained from the powder and the coating is presented in Fig. 5.8. The XRD analysis did not show the Co phase, as in XRD the cobalt remained undetected. Further, the SEM micrography, X-ray mapping and hardness testing confirm that pure Co is one phase. Co_3W_3C is formed because of the low carbon content (4 per cent) in the powder. The decomposition of WC was appreciable (I_{WC}/I_{W2C} is medium) in the coating but decarburization was present. This is due to the slow heat-up

Figure 5.8 X-ray diffraction of the powder and HVOF coating from two different samples, sprayed under similar conditions

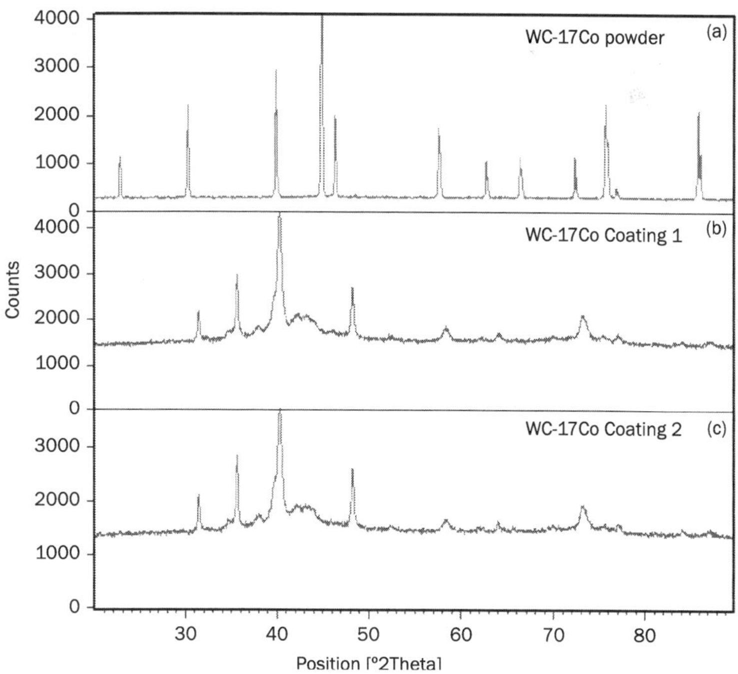

of WC particles and the presence of unavoidable oxygen in the HVOF process. Figure 5.8 also shows the consistency of the XRD results as analysed from randomly selected samples. The raised X-ray peaks from 34° to 48° indicate the presence of glassy Co-W-C phases in the coating.

A comparison of the XRD data as obtained from the HVOF and plasma coatings for the WC-17%Co powder is presented in Fig. 5.9. It is observed that under identical analysis conditions the intensities for $W_2C_{0.85}$ and elemental W are much higher in the optimized HVOF coatings than in the optimized plasma coatings. Similarly, it is noted that the degree of amorphous phase formation under optimized coating conditions is more significant in the case of HVOF-spayed WC-Co. This is obvious, as in addition to rapid quenching, available reaction times for formation of Co-W-C, stress and velocity factors are substantially higher in the case of WC-Co by HVOF with respect to those of WC-Co by plasma process. This confirms that WC-Co sprayed in air by either process undergoes the following decomposition reactions:

$$2WC \text{ (Sprayed in presence of air)} \longrightarrow W_2C_{0.85} + 1.15\ CO\text{(gas)} \qquad [5.3]$$

$$WC \text{ (Sprayed in presence of air)} \longrightarrow W_5 + CO\text{(gas)} \qquad [5.4]$$

The extent of formation of $W_2C_{0.85}$ and W is process dependent, especially in the case of a low temperature and high particle velocity process like HVOF.

The XRD results of WC-17%Co powder have also been considered for comparison purposes. All coatings on the samples showed the peaks of WC and less detectable Co phase, like the feedstock powder for all the Co contents, but a small peak of W and W_2C_x was observed in certain areas. This indicated that composition and phase changes took

Figure 5.9 X-ray diffraction of the HVOF coating and plasma coating

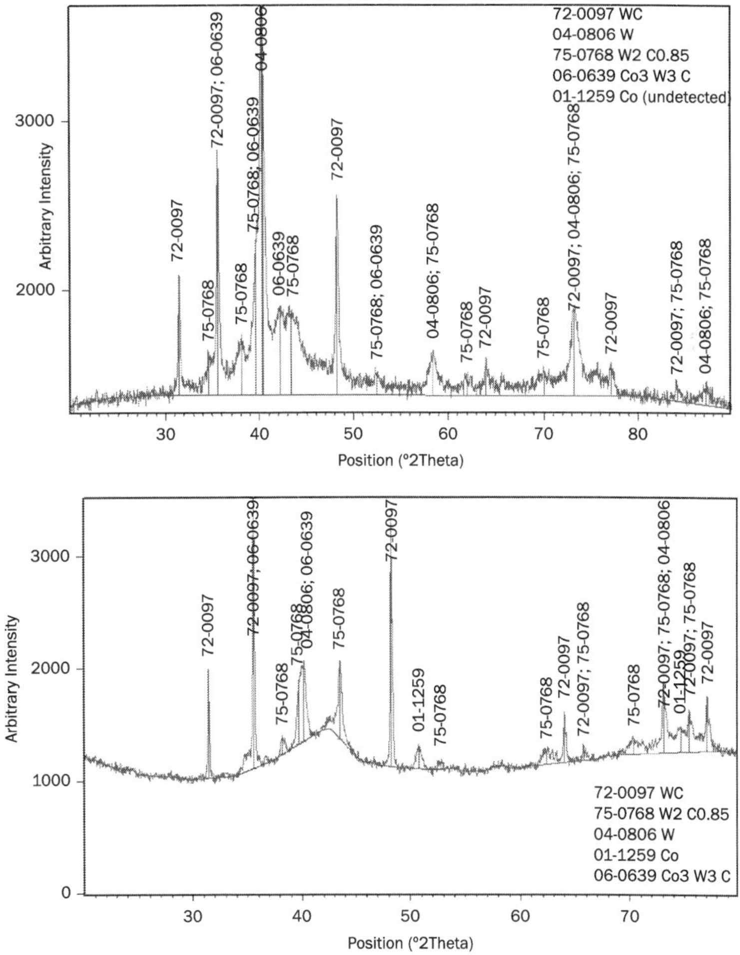

place during the deposition. Moreover, there are two points to be noted. The first is that WC peaks of the coatings are broadened compared to those of the powder. The second is that Co peaks of WC coatings are also broadened and shifted to lower angles. Possible explanations for these findings are the deformation of WC and Co phases due to high-velocity impact of powder during deposition, and dissolution of WC into the Co phase or decarburization of WC into W. The higher hardness value of WC-17%Co coating by plasma indicates that the matrix is harder compared to that coated by the HVOF process.

5.4 Tribological performances of WC-Co coatings

The wear resistance and other engineering properties of WC-Co coatings strongly depend on the spraying technique. Typically, as per ASTM G 65 standard testing, the abrasive wear rate of a quality WC-Co coating can be in a range from 0.9 to 16 mm^3 per thousand revolutions, while using the ASTM G 75 standard the erosive wear rate is somewhere between 1.0 and 5.0 μm per gram. The friction coefficients are around 0.18. The superior wear resistance of WC-Co is determined by evenly distributed fine tungsten carbide particles bonded to the cobalt matrix (Suzuki, 1986). In the case of the WC-Co coating the additional parameter is the adhesion of the coating to the substrate, as varying the thermal processes can result in varying heat transfer as well as overall adhesion characteristics. The basic schematic of such processes is illustrated in Fig. 5.10. The abrasion resistance of these coatings is found to be comparable to sintered cermets of the same composition. Mostly, the size effect of the abradant relative to the microstructure and splat size is important, as well as the innate inclination of the obtained phases to passivate and to control corrosion levels (Robert, 2010). The grain size of WC, its volume fraction, the relative amount of binder phase and the generated porosity determine the surface protection characteristics and fracture toughness of the WC-CO coatings. Since surface properties determine the wear

Figure 5.10 Graphical presentation of liquid droplet and substrate interaction in thermal spray process

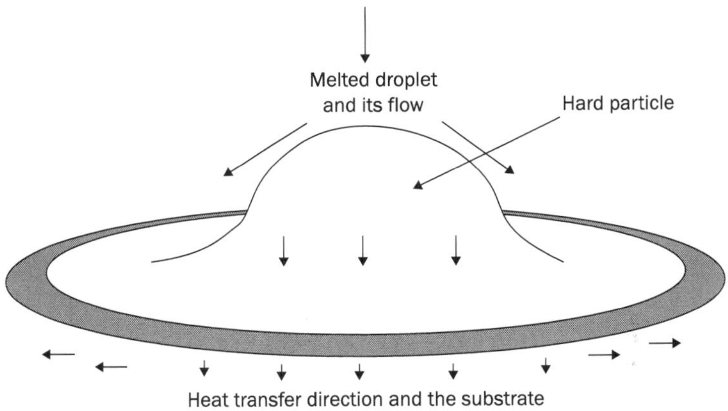

resistance of the components, the use of thermal spraying techniques has several attractive features for producing wear-resistant components. Wear performances under different wearing conditions are discussed below.

5.4.1 Sliding wear

Industrial applications where WC-Co coatings are used normally involve plasticity-dominated wear. Wear in this regime is governed by the deformation behaviour of the material undergoing wear. Material loss is primarily by delamination. The wear rate (W) in this regime is given by:

$$W = k\frac{F}{H} \quad [5.5]$$

where k is the known wear coefficient, F is applied load and H is hardness of wear of the material. The wear rate in this regime can be reduced by modifying the surface of a component with a layer having a) low wear coefficient k, or b) high hardness H.

Thermal sprayed WC-Co coatings are extensively used to improve the sliding wear properties of various engineering components, as these coatings have very high hardness and a reasonably low wear coefficient. An improvement of wear resistance of mild steel by 200 times by depositing detonation-sprayed WC-12%Co coating is reported by Roy and Sundararajan (2000). Their work is presented in Fig. 5.11. The friction coefficient of WC-Co coatings not only depends on the composition of the coatings but also on the deposition process. According to Barbezat et al. (1993), the friction coefficient of a WC-Co coating decreases with the increase in bond strength. Slavin and Nerz (1990) reported that friction coefficient of HVOF sprayed coating is lower than that of high energy plasma sprayed coating at different loads.

Kim et al. (2005) reported that the average friction coefficient of plasma-sprayed WC-12%Co coating decreases with the increase in bond strength or, in other words, the increase in inter-splat adhesive strength. Wear performance of WC-Co coatings is reported to significantly improve with heat treatment. Studies by Lenling et al. (1990) showed that post heat treatment of WC base coatings increased the compressive residual stress of these coatings. But the study by Stewart et al. (1998) showed that heat treatment reduced residual stress of the coating (Ito, 1990). An increase of residual stress in thermal spray coatings is very harmful as it may damage the coatings by spalling. On the whole, residual stress is one of the major problems in thermal spray layers, especially in thick coatings. Khameneh et al. (2004, 2006) also found deterioration of tribological properties of WC-Co coatings on heat treatment.

Nowadays, WC-Co-Cr powders are employed to improve wear and corrosion resistance (Berget, 2007; Chivavibul, 2007; Lee et al., 2010; Maiti, 2007; Murthy, 2001). Adhesion of the WC particles with the matrix (Karimi et al., 1993) is

Figure 5.11 Variation of wear rate as function of applied load of various modified surfaces (Roy and Sundararajan, 2000)

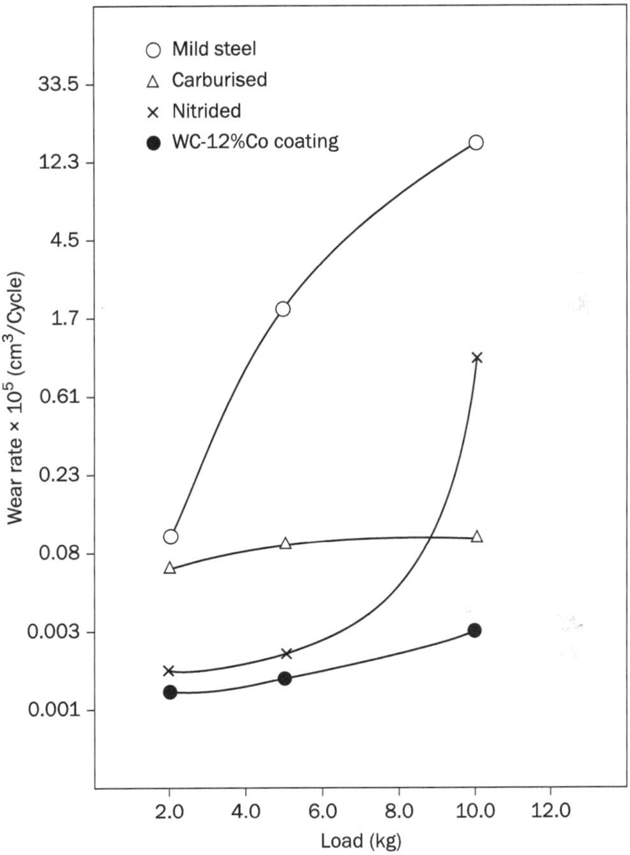

improved by adding Cr to Co. Lee et al. (2010) noted that the mixing of fine and coarse powders results in improvement of mechanical and tribological properties as shown in Fig. 5.12. It indicates best wear resistance for mixed powder containing 70 per cent fine and 30 per cent coarse powders. This behaviour can be explained by the fact that spaces generated by insufficient melting of coarse powders are filled by large amount of molten fine powders.

Figure 5.12 Bar diagram showing the influence of coating powder mixing on the wear rate of WC-Co coating (Karimi et al., 1993)

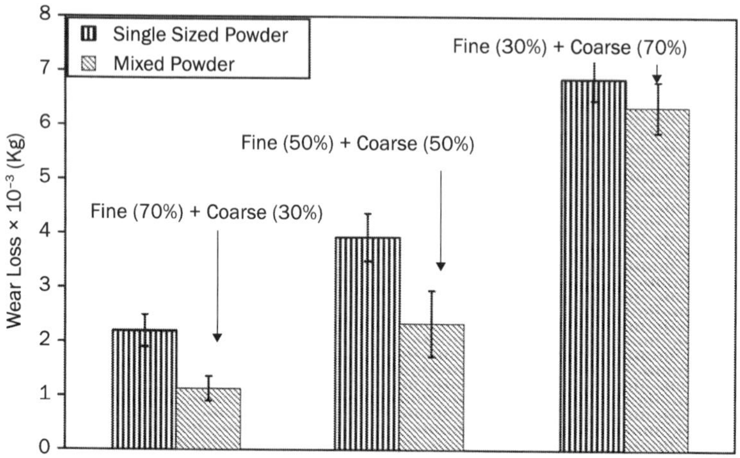

The nature of the fuel used also reflects wear performances. Work by Sudaprasert et al. (2003) indicates that wear rate of WC-12% coating increases as oxy gaseous fuel is changed to oxy liquid fuel. Their observation is presented in Fig. 5.13. During spraying with gaseous fuels, the binder material is totally molten and carbide dissolution is extensive, forming a highly alloyed matrix. On impacting, the binder solidifies leaving well-bonded carbides with the matrix phase. In contrast, when liquid fuel is used for spraying, the binder is partially molten, leaving a solid core and liquid shell. On impact, the solid core is damaged. This leads to poor wear resistance.

In the last few decades, nanostructured materials have received considerable attention due to their excellent properties (Brus, 1986; Ponce, 1997). However, difficulties in processing of such materials restricted their application (Morris, 1998; Wang, 1998). On the contrary, it is relatively easy to process nanocomposite coatings (Steward, 2000; Zhu

Figure 5.13 Bar diagram showing the influence of gaseous fuel and solid fuel on the wear rate of WC-Co coatings (Sudaprasert et al., 2003)

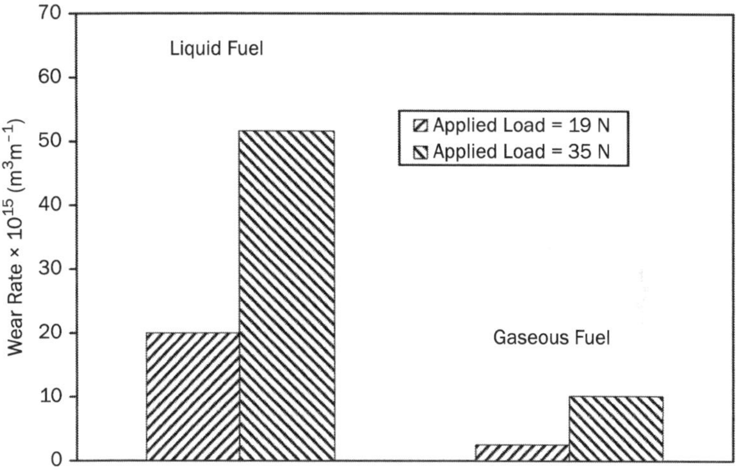

et al., 2001; Zhao et al., 2006; Baik et al., 2007). Significant improvements of wear performances of nanocomposite coatings have been reported. The work of Zhu et al. (2001), as illustrated in Fig. 5.14, indicates that the wear resistance of nanostructure WC-Co coatings is superior to conventional WC-Co coatings, particularly under high load. Such behaviour is attributed to higher hardness and toughness of nanostructure coatings as compared to conventional coatings. The wear in nanostructure coatings takes place by plastic deformation and mild surface fracture. In contrast, in conventional coating wear is due to removal of the binder phase followed by fragmentation and uprooting of carbide grains. A similar observation is also reported by Zhao et al. (2006) for plasma sprayed coatings. Baik et al. (2007) noted that coating with nanostructure WC-Co grains results in improvement of wear resistance because of retention of WC phases.

Figure 5.14 Variation of wear rate of nanocomposite coatings and conventional coatings as a function of applied load (Zhu et al., 2001)

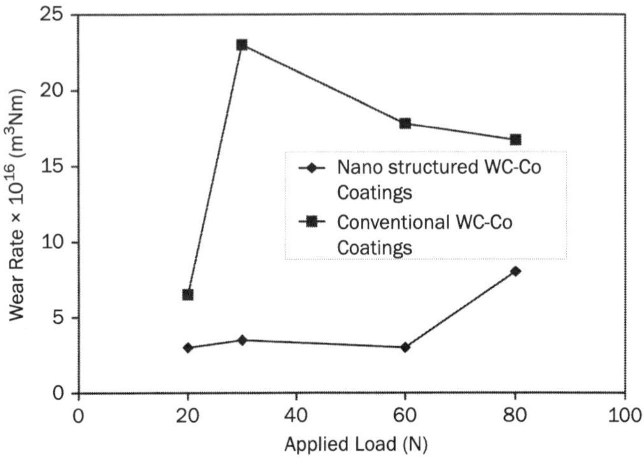

5.4.2 Erosive wear

Thermal sprayed coatings are also extensively used for protection against erosive wear. WC-Co has potential for use in erosion resistant coatings. Levy and Wang (1998) noted brittle erosion response of plasma sprayed WC-Co coating. Barbezat et al. (1993) observed that CDS-sprayed WC-Co coating exhibits improved erosion resistance. Impact angle independent erosion rates at low impact velocity and impact angle dependent erosion rates at high impact velocity for WC-Co coatings are noted by Wood et al. (1997). As reported by Karimi et al. (1993), the addition of Cr to Co results in improvement of adhesion of WC particles, with the matrix resulting in improved erosion resistance. According to Kim et al. (1994), the erosion rate of plasma-sprayed WC-Co coating decreases with increase of cohesive strength. Roy et al. (1993) demonstrated that the erosion resistance of detonation-sprayed WC-Co coating is better than plasma-sprayed and

Figure 5.15 Bar diagram showing erosion rate of various thermal sprayed coatings (Kulu et al., 2005)

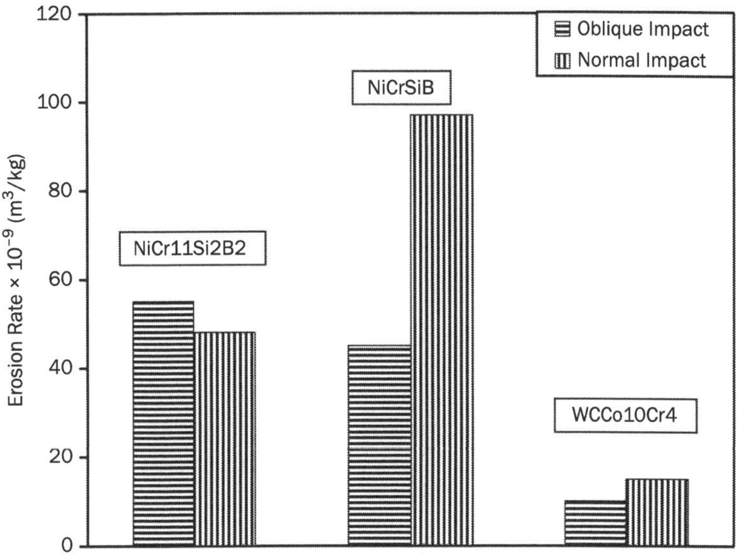

HVOF-sprayed coatings at normal impact, and comparable at oblique impact. The investigation by Kulu et al. (2005) presented in Fig. 5.15 indicates that WC-Co-Cr coating has significantly lower erosion rate than self-fluxing NI-Cr-Si-B coating. Unlike sliding wear, there is less work on solid particle erosion of nanocomposite coatings. Only work by Dent et al. (2002) indicates that erosion resistance of nanocomposite coating decreases compared to conventional coatings. However, in nanostructure coatings, erosion resistance increases with the decrease in Co binder phase.

5.4.3 Abrasive wear

According to Liu et al. (2003) the abrasive wear resistance of thermal spray coating can be empirically related to hardness and indentation toughness of the coating as given below:

$$WR_{abrasion} = C\frac{H^{1/2}K^{2/3}}{(1+nP)} \qquad [5.6]$$

where H is hardness, K is indentation toughness, P is porosity, n is a parameter to be determined experimentally and C is a constant. In general, most coatings and materials exhibited improved abrasion resistance with increase of hardness and toughness. Wang et al. (2009) were successful in increasing the hardness of WC-Co coating using a feedstock powder consisting of aggregates of hard ceramic particles that are either mixed or coated with more readily fusible nanophase binders, known as a multimodal powder. Increase in hardness resulted in improvement of abrasive wear resistance of WC-Co coatings. Their results are given in Fig. 5.16. It can be seen that abrasive wear resistance is directly related to the hardness of the coating, and toughness and porosity have minimum influence. Work by Kim et al. (1994) exhibited a decrease of abrasive wear rate with increase of bond strength.

Figure 5.16 Bar diagram showing abrasive wear rate of multimodal, conventional WC-Co coating along with the steel substrate and hard chrome plating (Wang et al., 2009)

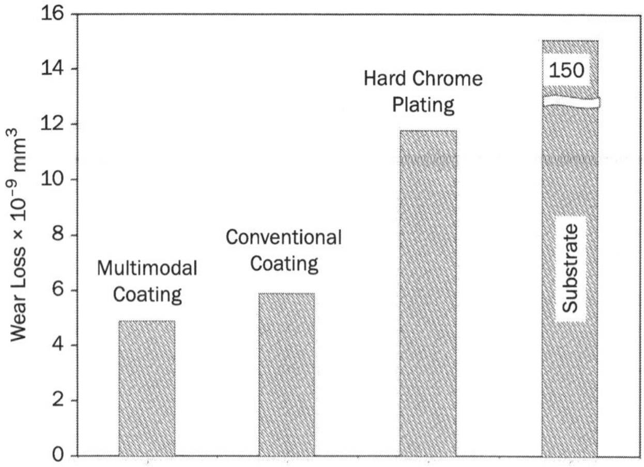

A systematic study by Barbezat et al. (1993) indicated that under three-body abrasion a WC-12%Co coating has a lower abrasion rate than a WC-17%Co coating. For WC-12%Co coatings, abrasion resistance is maximum for coating obtained by continuous detonation spraying, whereas for WC-17%Co coating abrasion is maximum for coating obtained by cyclic detonation spraying. It must be stated that minimum abrasion resistance can be seen for coatings obtained by vacuum plasma spraying for both varieties of coating. This observation was attributed to inter-splat adhesion strength. Nerz et al. (1991) demonstrated that abrasive wear resistance of high-energy plasma sprayed or HVOF-sprayed WC-Co coating increases when the coating is heat treated above recrystallization temperature due to formation of various hard carbides. An interesting study was reported by Kumari et al. (2010). According to their observation as shown in Fig. 5.17, the abrasive wear rate of WC-10Co-4Cr coating deposited by a JP 5000 gun increases with increase of mean free path of the

Figure 5.17 Variation of abrasive wear rate as function of binder mean free path (Kumari et al., 2010)

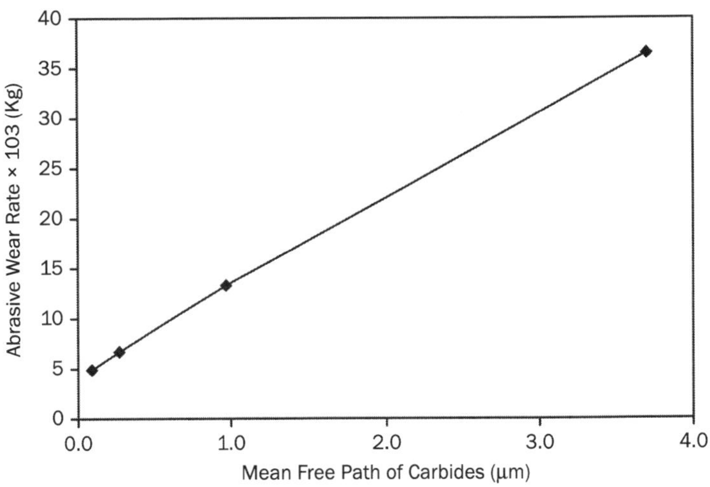

carbides. As materials loss takes place primarily by abrasion of binder followed by pulling out of carbide particle, binder abrasion becomes difficult due to small mean free path as the abrasive directly interacts with hard carbides.

Nanocomposite WC-Co coatings have also warranted attention for abrasive wear related applications. However, unlike sliding wear, nanocomposite coatings exhibited lower abrasion resistance than conventional coatings (Stewart, 1999) due primarily to the decarburization of wear-resistant WC particles and formation of amorphous phases. Similar behaviour is also noted by Dent et al. (2002), although they found that abrasive wear resistance increases with a decrease in Co binder content.

5.5 Applications of WC-Co coatings

Jet engines of fighter aircraft are equipped with fans and compressors. The vibrations of the blades of fans and compressor blades are controlled by midspan dampers, which are subjected to wear and tear. HVOF-sprayed WC-Co coatings are deposited on dampers to improve operating life (McGrann, 2000).

Thermal-sprayed WC-Co coating is extensively used in several places in the steel industries. Continuous casting moulds are coated with WC-Co coating. The top of the mould is coated with a thin WC-Co coating which can withstand high thermal stress in the zone of the molten steel meniscus. At the bottom, a thick coating which can withstand high ferrostatic pressure and thereby prevent cracking and spalling of the coating (Lavin, 1998) is used.

Briddle rolls are used to control the tension of steel strips as they passes through the continuous pickling, annealing and galvanizing lines. These rolls are conventionally coated with Cr plating and are increasingly being replaced by HVOF-sprayed WC-Co coatings due to their superior wear resistance (Sato, 1993). Similar results are obtained from detonation-sprayed WC-Co coatings against Cr plated deflector roll (Kasao, 1987).

WC-(10-15)%Co thermal-sprayed coatings are applied to galvanizing hardware due to their superior liquid metal erosion resistance relative to Fe and cobalt alloy (Ren, 2007; Sawa, 1995; Seong, 2000). However, success of this coating under such circumstances depends on the reaction between WC and Co phases to improve the poor corrosion properties of Co in molten Zn.

Murakawa and Watanabe (1989) demonstrated that an ironing die when coated with plasma-sprayed WC-Co coatings followed by vacuum sintering and hipping not only resulted in improved life of the die and increased output, but also in an improved surface finish of the product.

Figure 5.18 shows an application of WC-Co coating in an aircraft engine. The internal coating of stage III disc of an

Figure 5.18 Internal WC-17%Co coating in disc stage III of an aircraft engine

aircraft engine is illustrated in this figure. With this coating, disc life is reported to increase substantially. There are of course numerous other applications where WC-Co coatings are applied to protect components from tribological degradation.

5.6 Future trends

One important recent development in thermal spraying is the cold spray process, which can best be achieved by using expensive helium gases. Improvements in the He recovery process have, however, made this process quite competitive. Further development in nozzle design, powder quality and process optimization will result in further improved coating performances.

In recent decades various sensors relevant for thermal spraying process and capable of operating in the harsh environment of a spray booth have been developed (Fauchais, 2010). Today, sensors are available to measure trajectories, temperatures, velocities, sizes and shapes of in-flight particles (Fauchais, 1992; Gougeon, 1993; Landes, 2006; Li, 2003; Marple, 2001; Planche, 2004). Infrared cameras and pyrometers are employed to understand the temperature profile of substrate and coatings during preheating, spray process and cooling down (Clyne, 1996; Doubenskaia, 2006; Kuroda, 1988, 1991; Matejicek and Sampath, 2003; Matejicek, 2003; Verdelle, 2002; Xiaa, 2009). Sensors have also been developed to measure the stresses within the coatings and evolution of thickness of the coating during spraying (Nadeau, 2006). Further development towards improved precision in measurement is expected. New techniques such as shadowgraphs and lasers allow precise measurement of particle diameter not hitherto possible.

There should be a more comprehensive database on the tribological properties of a larger variety of compositions of WC-Co coatings. Further efforts should be directed towards the generation of data on high-temperature wear resistance of WC-Co coatings and wear behaviour of these coatings in the presence of corrosive fluids.

5.7 References

Baik K H, Kim J H and Seong B G (2007), 'Effects of post-spraying heat treatment on wear resistance of WC-Co nanocomposite coatings', *Mater. Sci. Eng. A*, **449–51**: 876.

Barbezat G, Nicoll A R and Sickinger A (1993), 'Abrasion, erosion and scuffing resistance of carbide and oxide ceramic thermal sprayed coatings for different applications', *Wear*, **162–4**: 529–37.

Barrella D J W and Priesta M (2003), 'The interaction of wear rate and friction with surface roughness for a lubricated sliding contact', *Tribology and Interface Engineering Series*, **43**: 807–14.

Berget J, Rohne T and Bardal E (2007), 'Erosion-corrosion properties of different WC-Co-Cr coatings deposited by the HVOF process—influence of metallic matrix composition and spray powder size distribution', *Surf. and Coat. Technol.*, **201**: 7619.

Bin Y, Yulong A, Huidi Z, Fengyuan Y and Jianmin C (2010), 'Friction and wear behaviours of plasma sprayed conventional and nanostructured WC-12Co coatings at elevated temperature', *Advanced Tribology, Part 3 (IV)*, 621–2.

Brus L (1986), 'Electronic wave functions in semiconductor clusters: experiment and theory', *J. Phy. Chem*, **90**: 2555.

Chivavibul P, Watanabe M, Kuroda S and Shinoda K (2007), 'Effects of carbide size and Co content on the microstructure and mechanical properties of HVOF-sprayed WC-Co coatings', *Surf. and Coat. Technol.*, **202**: 509.

Chun Hua H, Qiu Pu L, Feng J, Shi Ning M, Yu Lin Q et al. (2010), 'Tribological performances and friction reduction mechanism of N-Al_2O_3/FeS solid lubrication duplex layer', *Advanced Materials Research*, **146–7**: 243–7.

Clyne T W and Gill, S C (1996), 'Residual stresses in thermal spray coatings and their effect on interfacial adhesion: A review of recent work', *J. Thermal Spray Technol.*, **5**: 401.

Dent A H, DePalo S and Sampath S (2002), 'Examination of the wear properties of HVOF sprayed nanostructured and conventional WC-Co cermets with different binder phase contents', *J. Thermal Spray Technol.*, **11**: 551.

Doubenskaia M, Bertrand P and Smurov I (2006), 'Pyrometry in laser surface treatment', *Surf. Coat. Technol.*, **201**: 1955.

Fauchais P, Coudert J F, Vardelle M, Vardelle A, Denoirjean A (1992), 'Diagnostics of thermal spraying plasma jets', *J. Thermal Spray Technol.*, **1**: 117.

Fauchais P and Verdelle M (2010), 'Sensors in spray processes', *J. Thermal Spray Technol.*, **19**: 668.

Gao P, Yang G and Li C (2008), 'Preparation of multimodal structured WC-12Co deposits by cold spraying', *International Thermal Spray Conference & Exposition 2008: Thermal Spray Crossing Borders*, pp 1202–7. DVS-ASM.

Gougeon P and Moreau C (1993), 'In-flight particle surface temperature measurement: influence of the plasma light scattered by the particles', *J. Thermal Spray Technol.*, **2**: 229.

Ichiro M and Shigeyuki M (2007), 'Concept of molecular design towards additive technology for advanced lubricants', *Lubrication Science*, **19**(2): 127–49.

Ihn-Sung C, Seok-Hyung O and Jae-Youn J (2001), 'Lubrication characteristics between the vane and the rolling piston in a rotary compressor used for refrigeration and air-conditioning systems', *Journal of Mechanical Science and Technology*, **15**(5): 562–8.

Ito H, Nakamura R, Shiroyama M and Sasaki T (1990), 'Post-treatment of plasma sprayed WC-Co coatings by hot isostatic pressing', *Proc. Third Thermal Spray Conference*, p. 223. Long Beach CA, USA.

Karimi A, Verdon Ch and Barbezat G (1993), 'Microstructure and hydroabrasive wear behaviour of high velocity oxy-fuel thermally sprayed WC-Co(Cr) coatings', *Surf. and Coat. Technol.*, **57**: 81.

Kasai S, Sato Y, Yanagisawa A, Ichihara A and Onishi H (1987), 'Development of surface treatment techniques for process rolls in steelworks', *Report No. 17*, 81–90.

Khameneh A, Hyderzadeh M S, Hokamoto K and Umera M (2006), 'The effect of the heat treatment on residual stresses in HVOF sprayed WC-Co coating', *Wear*, **260**: 1203.

Khameneh A, Hyderzadeh M S and Hadavi, S M M (2004), 'Wear and erosion behavior of plasma-sprayed WC-Co coatings', *Mater. Sc. Forum*, **427**: 465–6.

Kim H J, Kweon Y G and Chang R W (1994), 'Wear in relation to friction – a review', *J. Thermal Spraying*, **3**: 169.

Koji K (2000), 'Solid particle erosion of thermal sprayed coatings', *Wear*, **241**(2): 151.

Kulu P, Hussainova I and Veinthal R (2005), 'Effect of microstructure on abrasive wear behavior of thermally sprayed WC-10Co-4Cr coatings', *Wear*, **258**: 488.

Kumari K, Anand K, Bellaci M and Giannozzi M (2010), 'The Quenching Stress in Thermally Sprayed Coatings', *Wear*, **268**: 1309.

Kuroda S and Clyne TW (1991), 'Simultaneous measurement of coating thickness and deposition stress during thermal spraying', *Thin Solid Films*, **200**: 49.

Kuroda S, Fukushima T and Kitahara S (1988), 'Diagnostics in plasma spraying techniques', *Thin Solid Films*, **164**: 157.

Landes K (2006), 'A study on powder mixing for high fracture toughness and wear resistance of WC-Co-Cr coatings sprayed by HVOF, Surface and coatings technology', *Surf. Coat. Technol.*, **201**: 1948.

Lavin P (1998), 'Coating of continuous casting machine components', International Patent, Publication Number WO 98/21379. Monitor Coating and Engineering Limited.

Lee C H, Han J H, Yoon J, Shin M C and Kwun S I (2010), 'Heat treatment of thermally sprayed WC-Co coatings', *Surf. and Coat. Technol.*, **204**: 2223.

Lenling W J, Smith M F and Henfling J A (1990), 'Erosion of hard material coating systems', *Proc. Third Thermal Spray Conference,* p. 227. Long Beach CA, USA.

Levy A V and Wang B (1988), 'Effect of spray particle trajectory on the measurement signal of particle parameters based on thermal radiation', *Wear*, **121**: 325.

Li C J, Wu T, Li C-X and Sun B (2003) 'Comparison of HVOF and plasma-sprayed alumina/titania coatings-microstructure, mechanical properties and abrasion behavior', *J. Thermal Spray Technol.*, **12**: 80.

Liu Y, Fischer T and Dent A (2003), 'Improving the wear behavior of WC-CoCr-based HVOF coating by surface grinding', *Surf. and Coat. Technol.*, **167**: 68.

Mainsah E, Greenwood J A and Chetwynd DE (2001), *Metrology and Properties of Engineering Surfaces*, London: Springer Verlag.

Maiti A K, Mukhopadhyah N and Raman R (2007), 'Thermal spraying of nanostructured cermet coatings', *Surf. and Coat. Technol.*, **201**: 7781.

Marple B R, Voyer J, Bisson J F and Moreau C (2001), 'In situ measurement of residual stresses and elastic moduli in thermal sprayed coatings (Part 1): Apparatus and analysis, Acta Materilla, 51:863', *J. Mater. Process. Technol.*, **117**: 418.

Matejicek J and Sampath S (2003), 'In situ measurement of residual stresses and elastic moduli in thermal sprayed coatings (Part 2): Processing effects on properties of Mo coatings', *Acta Meter.*, **51**: 863.

Matejicek J, Sampath S, Gilmore D and Neiser R (2003), 'Fatigue life in bending and coating residual stress in tungsten carbide thermal spray coatings', *Acta Meter.*, **51**: 873.

McCool J I (1986), 'Comparison of models for the contact of rough surfaces', *Wear*, **107**: 37–60.

McGrann R T R and Shanley J R (2000), in C C Berndt (Ed.) *Thermal Spray: a United Forum for Scientific and Technological Advances*, pp. 341–9. Ohio: ASM International.

Michael N (2003), 'Tribology in design', *Tribology and Interface Engineering Series*, 3–11.

Mohanty R M (2009), 'Powder application in surface engineering', Powder metallurgy short course ISSN-0377-8452, August 17–21. Mumbai: Powder Metallurgy Association of India.

Morris D G and Morris (1998),'Wear mapping and wear characterization methodology', *M A, Mater. Sci. Eng A*, **104**: 201.

Murakawa M and Watanabe S (1989), 'Effect of grinding on the erosion behaviour of a WC-Co-Cr coating deposited by HVOF and detonation gun spray processes', *Proc. 2nd Intl. Conf. on Hot Isostatic Processing, Theory and Applications*, 7–9 June, Gaitherburg, MD, USA.

Murthy K N, Rao D S and Venkataraman D (2001), 'A new approach to online thickness measurement of thermal spray coatings', *Wear*, **249**: 592.

Nadeau A, Pouliot L, Nadeau F, Blain J, Berube S A et al. (2006), 'Micro-structural evaluation of tungsten carbide coatings', *J. Thermal Spray Technol.*, **15**: 744.

Nakada M (1995), 'Piston and piston ring tribology and fuel economy', *Proceedings of International Tribology Conference*, Yokohama.

Nerz J E, Kushner B A, Rotolico A J (1991), 'Evaluation of KIC of brittle solids by the indentation method with low crack-to-indentation ratios', *Proc. 4th National Thermal Spraying Conference*, 4–10 May, Pittsburg, PA, USA.

Niihara K, Morena R and Hassleman D P H (1982), 'Fracture mechanics of ceramics', *J. Mater. Sci. Lett*, **1**: 13–16.

Niihara K, Morena R and Hassleman D P H (1983), in R C Bradt, A G Evans, D P Hassleman and F F Lange (eds) *Fracture Mechanics of Ceramics*, pp. 97–105, New York: Plenum Press.

Planche M P, Liao H and Coddet C (2004), 'Relationships between in flight particle characterstics and Coating Microstructure with a twin wire arc spray processes and different working conditions', *Surf. Coat. Technol*, **182**: 215.

Ponce F A and Bour, D P (1997), 'Semiconductors used for light emitting devices', *Nature*, **386**: 351.

Powlowski L (1995), *The Science and Engineering of Thermal Spray Coatings*, New York: John Wiley.

Ram B B, Joseph C C, Maurice F A and Brezler R A (1996), 'Tribological performance evaluation of tungsten carbide-based cermets and development of a fracture mechanics wear model', *Wear*, **201**(1–2): 233–43.

Ren X, Mei X, She J and Ma J (2007), 'Materials resistance to liquid zinc corrosion of surface of sink roll', *Proc. of Sino-Swedish Structural Materials Symposium*, 125–31.

Robert J K (2010), 'Tribology of thermal sprayed WC-Co coatings', *Journal of Refractory Metals and Hard Materials*, **28**(1): 82–94.

Roy M (2002), 'Dynamic hardness of detonation sprayed WC-Co coatings', *J. Thermal Spraying*, **11**: 393–9.

Roy M (2005), 'Application of x-ray technique to surface modification: case studies', in A K Singh (ed.) *Advanced X-ray Techniques in Research and Industry*, pp. 416–24, Capital Publishing Company, New Delhi.

Roy M (2006), 'Mode of deformation during tribological degradation', *Proc. Intl. Conf. on Advances in Materials and Materials Processing*, Kharagpur, India, 202–9.

Roy M and Sundararajan G (2000), 'The reciprocating wear of detonation sprayed WC-Co coating on mild steel', in W J Bartz (ed.) *Tribology 2000 – Plus*, pp. 1971–9, Stuttgart: Technische Akademie Esslingen.

Roy M, Narkhede B E and Paul S N (1999), 'Tribological performances of WC-Co coating', in H Prasad (ed.) *Tribology in 2000 and Beyond*, pp. 205–15, Hyderabad: Kalajyothi Process Ltd.

Salimijazi H R, Pershin L, Coyle T W, Mostaghimi J, Chandra S et al. (2007),' Measuring substrate temperature variation during application of plasma sprayed zirconia coatings', *J. of Thermal Spray Technol.*, **16**: 580.

Salvin T P and Nerz J (1990), 'Thermal spray research and applications', *Proc. 3rd National. Thermal Spraying Conf.*, Long Beach, CA, USA, 20–25 May.

Sato Y, Midorikawa S, Iwashita Y, Yokogawa A and Takano T (1993), 'Service life extension technique for cold rolling rolls', *Kawasaki Steel Technical Report No. 29*, 74–82.

Sawa M and Oohori J (1995), 'Application of thermal spraying technology at steelworks', in A. Ohmri (ed.) 'Thermal Spraying: Current Status and Future Tends', *Proc. 14th International Thermal Spray Conference*, May 22–26, Kobe, Japan, pp. 37–42. Kobe: High Temperature Society of Japan.

Seong B G, Hwang S Y, Kim M C and Kimin K Y (2000), 'Observation on the WC-Co coating used in a zinc pot of a continuous galvanizing line', in C C Berndt (ed.) *Thermal Spray: Surface Eng. via Applied Research*, pp. 1159–67, Ohio: ASM International.

Simon Tung (2008), *Automotive Lubricant Testing and Advanced Additives*, p. 143, Ohio: ASM International.

Sinha A K (2003), *Physical Metallurgy Handbook*, New York: McGraw-Hill.

Smart N and Bockris J (1987), 'Effect of water activity on corrosion', *Corrosion*, **48**: 277–80.

Stewart D A, Shipway P H and McCartney D G (2000), 'Microstructural evolution in thermally sprayed WC-Co coatings: comparison between nanocomposite and conventional starting powders', *Acta. Mater.*, **48**: 1593.

Stewart D A, Shipway P H and McCartney D G (1999), 'Influence of heat treatment on the abrasive wear behaviour of HVOF sprayed WC-Co coatings', *Wear*, **189**: 225–9.

Stewart D A, Shipway P and McCartney D G (1998), 'Influence of heat treatment on the abrasive wear behavior of HVOF sprayed WC-Co coatings', *Surf. Coat. Technol.*, **105**: 13.

Stewart D, Shipway P, McCartney D G (1998), 'Influence of heat treatment on the abrasive wear behaviour of HVOF sprayed WC-Co coatings', *Surface & Coatings Technology*, **105**(1–2): 13–24.

Sudaprasert T, Shipway P H and McCartney D G (2003), 'Sliding wear behaviour of HVOF sprayed WC-Co coatings deposited with both gas-fuelled and liquid-fuelled systems', *Wear*, **255**: 943.

Suzuki H (1986), *Cemented Carbides and Sintered Hard Metals*, Tokyo: Maruzen Co Ltd., translated.

Tabor D (1951), *The Hardness of Materials*, Oxford: Clarendon Press.

Verdelle M, Renault T and Fauchais P (2002), 'Choice of an IR pyrometer to measure the surface temperature of a coating during its formation in air plasma spraying', *High Temp Mater. Process*, **6**: 469.

Vinayo M E, Kassabji F, Guyonnet J and Fauchais P (1985), 'Plasma sprayed WC-Co coatings: Influence of spray conditions (atmospheric and low pressure plasma spraying) on the crystal structure, porosity, and hardness', *Journal of Vacuum Science & Technology A: Vacuum, Surfaces, and Films*, 3(6): 2483–9.

Voevodin A, O'Neill J and Zabinski J (1999), 'Tribologica, performance and tribochemistry of nanocrystalline WC/ amorphous diamond-like carbon composites', *Thin Solid Films*, **342**: 194–200.

Wang Q, Chen Z H and Ding, Z X (2009), 'Performance of abrasive wear of WC-12Co coatings sprayed by HVOF tribology international', *Tribology International*, **42**: 1046.

Wang S and Wei W J (1998), 'Characterization of Al2O3 composites with Mo particulates, II. densification and mechanical properties', *Nanostruct. Mater.*, **10**: 983.

Williams J A (2000), *Engineering Tribology*, Cambridge: Cambridge University Press.

Wood R J K, Mellor B G and Binfield M L (1997), 'Sand erosion performance of detonation gun applied tungsten carbide/cobalt-chromium coatings', *Wear*, **211**: 70.

Xiaa W, Zangb H, Wanga G and Wanga Y (2009), 'Novel integrated temperature investigation approach of sprayed coatings during APS process.

Zhao X Q, Zhou H D and Chen J M (2006), 'Comparative study of the friction and wear behavior of plasma sprayed conventional and nanostructured WC-12%Co coatings on stainless steel, Materials Science and Engineering', *Mater. Sci. Eng: A*, **431**: 290.

Zhu Y C, Yukimura K, Ding C X and Zhang P Y (2001), 'Tribological properties of nanostructured and conventional WC-Co coatings deposited by plasma spraying', *Thin Solid Films*, **388**: 277.

6

Tribological performance of electroless Ni-P coatings

P. Sahoo, Jadavpur University, India

Abstract: Electroless Ni-P coatings are quite suitable for tribology based applications. They are smooth due to the nodular surface and have good wear resistance. Hardness of Ni-P coating and wear resistance increase with heat treatment. Several particle inclusions in the coating are found to favor its tribological performance. Three tribological testing parameters – applied load, sliding speed and sliding duration – are optimized in order to minimize friction and wear of the coating. Taguchi method with L_{27} orthogonal array is employed to optimize the individual performances whereas grey relational analysis is considered for optimization of the combined friction and wear problem. Analysis of variance is used to detect the influential testing parameters and their interactions. All the testing parameters are found to have significant influence on the individual as well as the combined tribological performance of the coating.

Key words: electroless Ni-P, wear, friction, Taguchi, grey relational analysis.

6.1 Introduction

Tribology is the science of interacting surfaces in relative motion. Contacting surfaces in relative motion give rise to two undesirable phenomena: friction and wear. Development of newer materials is constantly attempted by researchers to combat friction and wear. The surface coating field is a rapidly developing area of science and technology that offers new methods and techniques to control friction and wear. Newer coating types are continually being developed and potential applications in different industrial fields are ever growing, ranging from machine components and consumer products to medical instruments and prostheses (Holmberg and Matthews, 1998). Electroless nickel coating is one such coating that has a wide range of applications.

Electroless Ni-P (EN) coatings, since their introduction in 1946 by Brenner and Riddell (Brenner and Riddell, 1950), have received wide acceptance in numerous industrial applications (Riedel, 1991) due to their excellent mechanical, physical, electrical, corrosion and wear resistance properties. They are commonly used in engineering coating applications where wear resistance, hardness and corrosion protection are of prime importance. Applications include oil field valves, rotors, drive shafts, paper handling equipment, fuel rails, optical surfaces for diamond turning, door knobs, kitchen utensils, bathroom fixtures, electrical/mechanical tools and office equipment. They are also commonly used as coatings in electronic printed circuit board manufacturing, typically with an overlay of gold to prevent corrosion; a process known as electroless nickel immersion in gold. Due to the high hardness of the coating it can be used to salvage parts. Coatings of 25 to 100 μm can be applied and machined back to final dimensions. They are also used extensively in the manufacture of hard disk drives, as a way of providing an atomically smooth coating to the aluminium disks present in the drive. EN coatings can be applied to a variety of substrate materials and plated uniformly on intricate part geometries.

EN coating is an autocatalytic deposition of a Ni-P alloy from an aqueous solution onto a substrate without the application of electric current. Thus, the electroless deposition process is different from the conventional electroplating processes that require an external source of direct current in order to reduce nickel ions in the electrolyte to nickel metal on the substrate. The electroless bath typically comprises an aqueous solution of metal ions, complexing agents, reducing agents and stabilizers, operating in a specific metal ion concentration, temperature and pH range. The deposition rate, properties of coated components and the structural behavior of deposits mainly depend on the plating bath constituents and conditions, such as the type and concentrations of the reducing agent, stabilizer, pH used and the temperature of the bath, etc.

Since the inception of EN coatings, the properties and structures of such coatings have received considerable research attention. The properties and microstructures of EN coatings depend on the amount of phosphorus alloyed in the deposit (Agarwala and Agarwala, 2003; Allen and Vandersande, 1982; Baudrand and Bengston, 1995; Berkh et al., 1996; Hur et al., 1990). The post-deposition heat treatment influences the properties by altering the microstructures (Agarwala and Agarwala, 2003; Apachitei et al., 1998; Baudrand and Bengston, 1995; Hur et al., 1990). The mechanical and tribological properties of these coatings can be further improved by the incorporation of hard particles (Apachitei et al., 1998; Huang et al., 2003) and dry lubricants (Huang et al., 2003; Straffelini et al., 1999; Zhao et al., 2002). The evolution of friction and wear has been monitored in order to detect the friction and wear transitions and therefore to determine the surface durability of the co-deposits (Straffelini et al., 1999). Friction and wear have been the focus of attention of a number of researchers (Alirezaei et al., 2007; Quercia et al., 2001; Ramalho and Miranda, 2005; Sahoo and Das, 2011; Staia et al., 1996 and 1997; Taheri et al., 2001) over the years.

6.2 Friction behavior of electroless Ni-P coatings

Electroless nickel coatings are in general very smooth and lubricious (Balaraju et al. 2003) in nature due to their unique nodular microstructure resembling that of a cauliflower (Delaunois and Lienard, 2002). But different surface treatments and incorporation of various particles and elements lead to a change of the conventional microstructure of the coating resulting in modified friction behavior. Moreover, the friction characteristics are also dependent on the condition of the tribological testing and the counter face material used. It is in general observed that the friction coefficient of EN coatings decreases with increase in load. The friction coefficient was found to be within the range of 0.15–0.35 when tested under the 15–60 N loading conditions (Staia et al., 1996). The friction study of EN coatings concluded that coatings with high phosphorus content have higher friction coefficient than medium or low phosphorus electroless coatings (Taheri et al., 2001). The friction coefficient of electroless coatings having 6–7 per cent phosphorus content when tested under low loads was found to be as high as 0.7 (Staia et al., 1996). Heat treatment in general results in a reduction of the friction coefficient of various electroless coatings compared to the as-deposited coatings. Amplitude of the friction coefficient oscillation of

Ni-P coating after tempering is found to be lower than that of non-tempered Ni-P coating (Li et al., 2008).

The addition of hard particles such as B_4C and SiC has a tendency to increase the friction coefficient of electroless nickel coatings as the natural lubricity of the coatings is lost. The friction coefficient of EN coatings is found to increase after the addition of B_4C particles (Araghi and Paydar, 2010; Ebrahimian-Hosseinabadi et al., 2006), which may be attributed to the particles of B_4C separated during testing and probably imprisoned between the specimen surface and the pin (Araghi and Paydar, 2010). The friction coefficient of electroless Ni-P-B_4C composite coatings with 25 vol.% of B_4C is approximately doubled when increasing the load from 15 to 60 N (Ebrahimian-Hosseinabadi et al., 2006). Lower friction coefficients may correspond to the presence of surface oxides formed as a result of frictional heating, which is more considerable for higher loads. The co-deposition of SiC particles in Ni-P coatings also increases the friction coefficient of the coating and Wu et al. (2006a) found an increase of about 10 per cent in friction coefficient when compared to pure Ni-P coatings. On the other hand, the friction coefficient showed a drastic reduction when soft particles such as PTFE, graphite and MoS_2 are introduced in the coating.

The incorporation of PTFE (Wu et al., 2006a) and graphite (Wu et al., 2006b) in Ni-P matrix decreased the friction coefficient markedly and facilitated the stable state of the whole wear course. In a study by Wu et al. (2006a), a decrement as large as 70 per cent was observed in the friction coefficient of Ni-P coating after the introduction of PTFE in its matrix. The PTFE-rich mechanically mixed layer (PRMML) formed on the worn surface is responsible for the good anti-friction properties.

The friction coefficient of the Ni-P-nano-MoS_2 composite coating decreased greatly compared with that of EN coatings

Figure 6.1 Variation of friction of Ni-P-IF MoS_2 with different volume fractions of IF MoS_2 at load of 15 N (Zou et al., 2006)

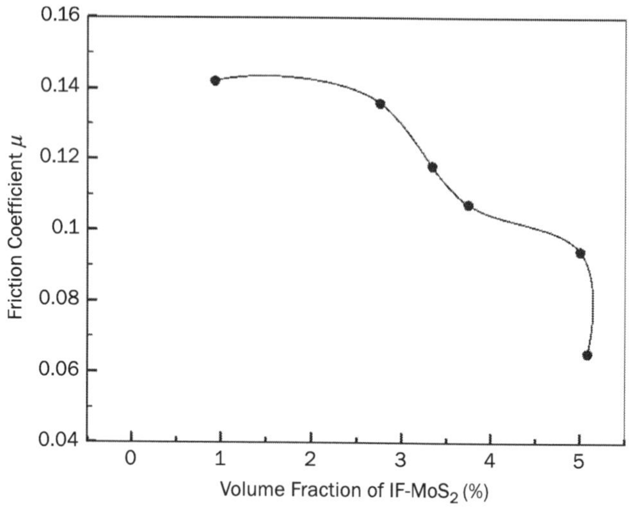

(Hu et al., 2009) and showed a decreasing trend with the increase of the volume fractions of IF-MoS_2 (Zou et al., 2006) (Fig. 6.1). Moreover, Ni-P-(IF-MoS_2) composite coating displayed excellent friction properties in a vacuum, implying the good stability of IF-MoS_2 in different environments. The favorable effects of IF-MoS_2 nanoparticles on the friction properties are attributed to their unique fullerene-like structure. Incorporation of SiO_2 nanoparticles also contributed to the improvement of the friction-reducing ability of the EN matrix (Dong et al., 2009).

6.3 Wear behavior of electroless Ni-P coatings

One of the unique characteristics of electroless nickel deposition is the superior wear resistance of the coatings.

Theoretically, there is a correlation between wear resistance and hardness of a surface. However, the wear properties of a surface are affected by numerous other parameters such as the nature of the applied stress and the surface morphology. The wear resistance of electroless nickel deposits depends on both phosphorus content and the type of post heat treatment applied. In general, heat treatment increases hardness and hence the wear resistance of the coating, but grain coarsening at higher heat treatment temperatures negatively affects the wear resistance of the coating. Laser irradiated coatings showed less wear than as-deposited amorphous alloy and furnace annealed coatings (Tsujikawa et al., 2005).

The wear mechanism in electroless nickel coating is primarily found to be either adhesive or abrasive, or a combination of both. The mechanism of wear of the EN deposit depends on the attractive force that operates between the atoms of nickel from the coating and iron from the counter disk. A negative wear depth curve is also observed in some cases (Palaniappa and Seshadri, 2008) as a result of the build-up of oxide debris at the interface of the coated pin and counter disc. Moreover, this build-up of oxides acts as a lubricant at the interface resulting in lower frictional coefficient in case of heat-treated coatings.

Co-deposition of both tungsten and ZrO_2 particles improves the wear resistance of the EN-based coatings further (Gay et al., 2007). Hamid et al. (2007) reported that co-depositing tungsten carbide (WC) with Ni-P matrix improved the abrasion resistance of steel 13 times while EN improved the abrasion resistance of steel 3 times. Higher abrasion resistance of Ni-P–WC coatings is attributed to increase of the effective role of the Ni matrix for supporting and keeping the particles in the matrix so that the coating can sustain shear stress, rupture and damage. A particular volume per cent (53) incorporation of WC was found to suffer the minimum weight loss in wear testing. Several other soft and hard particulates improve the wear performance of electroless nickel coating which include: alumina (Alirezaei et al., 2005; Madej et al., 2006; Novak et al., 2010), boron carbide (Araghi and Paydar, 2010; Ebrahimian-Hosseinabadi et al., 2006), silicon carbide (Sarret et al., 2006; Wu et al., 2006b), silicon nitride (Das et al., 2007), silicon oxide (Dong et al., 2009), PTFE (Ramalho and Miranda, 2005; Srinivasan and John, 2005; Wu et al., 2006a) and molybdenum sulphide (Zou et al., 2006).

Some enthusiasts have also studied the impact of various tribological testing conditions on the wear behavior of electroless nickel coatings. It is found that rate of wear is strongly dependent on the applied load and it increases along with increasing load (Krishnaveni et al., 2005) irrespective of whether specimen is Ni-P coated or Ni-P coated and tempered. But then load is found to have stronger impact on the wear rate for Ni-P coating, as Li et al. (2008) have observed that an increase of load from 1 N to 1.5 N magnifies the wear rate by almost 10 times; the same increase with tempered Ni-P coatings magnifies the wear rate by only 1.3.

6.4 Electroless nickel bath characteristics and role of individual components

Electroless plating is an autocatalytic process where the substrate develops a potential when it is dipped in an electroless solution, a bath that contains a source of metallic ions, reducing agent, complexant, stabilizer and other components. Due to the developed potential, both positive and negative ions are attracted towards the substrate surface and release their energy through the charge transfer process. The components of electroless nickel bath and their functions are given in Table 6.1.

6.4.1 Reducing agents

Two types of bath have been mainly used for depositing Ni-B alloy, acidic and alkaline baths depending on the reducing agent used, as listed in Table 6.2 (Delaunois et al., 2000). Hypophosphite baths are the most common type of

Table 6.1 Electroless nickel bath components and their functions

Component	Function
Nickel ion	Source of metal
Reducing agent	Source of electrons
Complexants	Stabilizes the solution
Accelerators	Activates reducing agent
Buffers	Controlling pH (long term)
pH regulators	Regulates pH of solution (short term)
Stabilizer	Prevents solution breakdown
Wetting agents	Increases wettability of the surfaces

Table 6.2 Reducing agents for EN plating

Deposit	Reducing agent	Remarks
Ni-P	Sodium hypophosphite	Acid or alkaline bath (2–17%)
Ni-B	Sodium borohydride ($NaBH_4$)	Acid or alkaline bath
	Aminoborane (DMAB)	Alkaline bath (0.5–10% B)
Only Ni	Hydrazine (NH_2NH_2)	Alkaline bath

commercially used electroless nickel bath due to higher deposition rates, increased stability and greater simplicity of bath control. The mechanism of the EN deposition reactions is still not well understood. However, the most widely accepted mechanisms are illustrated by the following equations (Agarwala and Agarwala, 2003):

1. Electrochemical mechanism, where catalytic oxidation of the hypophosphite yields electrons at the catalytic surface which in turn reduces nickel and hydrogen ions, is illustrated below:

$$H_2PO_2^- + H_2O \rightarrow H_2PO_3^- + 2H^+ + 2e^- \quad [6.1]$$

$$Ni^{++} + 2e^- \rightarrow Ni \quad [6.2]$$

$$2H^+ + 2e^- \rightarrow H_2 \quad [6.3]$$

$$H_2PO_2^- + 2H^+ + e^- \rightarrow P + 2H_2O \quad [6.4]$$

2. Atomic hydrogen mechanism, where atomic hydrogen is released as the result of the catalytic dehydrogenation of hypophosphite molecule adsorbed at the surface is illustrated below:

$$H_2PO_2^- + H_2O \rightarrow HPO_3^{--} + H^+ + 2H_{ads} \quad [6.5]$$

$$2H_{ads} + Ni^{++} \rightarrow Ni + 2\,H^+ \quad [6.6]$$

$$H_2PO_2 + H_{ads} \rightarrow H_2O + OH^- + P \quad [6.7]$$

The adsorbed active hydrogen, (6) reduces nickel at the surface of the catalyst.

$$(H_2PO_2)^{2-} + H_2O \rightarrow H^+ + (HPO_3)^{2-} + H_2 \quad [6.8]$$

Simultaneously, some of the absorbed hydrogen reduces a small amount of the hypophosphite at the catalytic surface to water, hydroxyl ion and phosphorus. Most of the hypophosphite present is catalytic, which is oxidized to orthophosphite and gaseous hydrogen, causing low efficiency of electroless nickel solutions for alloy coating while the deposition of nickel and phosphorus continues. Several other hypotheses regarding the reaction mechanism have been put forward by various researchers (Mallory and Hadju, 1991).

6.4.2 Complexing agents

There are three principal functions that complexing agents perform in the electroless nickel plating bath (Mallory and Hadju, 1991):

- They exert a buffering action that prevents the pH of the solution from falling too fast.
- They prevent the precipitation of nickel salts, e.g., basic salts or phosphites.
- They reduce the concentration of free nickel ions by forming meta-stable complexes.

Moreover, the complexing agent also influences the reaction mechanism and deposition rate and hence the deposit. Generally, all complexing agents follow an inverted bell shaped curve when their concentration is compared with the deposition rate, i.e. there is a certain concentration when the deposition rate is maximal. The deposition rate gradually increases up to that optimum concentration and then falls. In the case of reduction by hypophosphite, the optimum concentration for sodium citrate, which acts as the complexing agent, is about 30 g/l (Agarwala and Agarwala, 2003). The optimum concentration for ethylenediamine is found to be 90 g/l (Anik et al., 2008). Ammonium fluoride improves the deposition rate and the buffering capability of a Ni-P bath (Ying et al., 2007).

Generally, the complexing agents are made of organic acids or their salts, i.e. acetate, succinate, propionate, citrate, etc. The only inorganic compounds used as complexing agents are the pyrophosphate anion and the ammonium ion (Mallory and Hadju, 1991).

6.5 Optimization of tribological performance of electroless Ni-P coatings

From the previous discussions it can be seen that EN coatings are well suited for tribological applications. Hence, to evaluate their tribological performance in a more elaborate manner, an optimization scheme is used in which the tribological testing parameters are varied in order to obtain an enhanced tribological performance from the coating. Taguchi orthogonal array is used for the optimization along with grey relational analysis for multiple response optimizations. In the present case, friction and wear are considered as the tribological performance characteristics.

6.5.1 *Need for design of experiments*

Industrial physicists can no longer afford to experiment in a trial-and-error manner, changing one factor at a time, the way Edison did in developing the light bulb. A far more effective method is to apply a computer-enhanced, systematic approach to experimentation, one that considers all factors simultaneously. That approach is called design of experiments (DOE), and corporations worldwide are adopting it as a cost-effective way to solve serious problems afflicting their operations. DOE provides information about the interaction of factors and the way the total system works, something not obtainable through testing one factor at a time (OFAT) while holding other factors constant. Another advantage of DOE is that it shows how interconnected factors respond over a wide range of values, without requiring the testing of all possible values directly.

The concepts of DOE have been in use since Fisher's work in agricultural experimentation approximately half a century ago. Fisher (1951) successfully designed experiments to determine optimum treatments for land to achieve maximum yield. Numerous applications of this approach, especially in the chemical and pharmaceutical industries, are cited in the literature.

In general, a system or process (Fig. 6.2) can be visualized as a combination of machines, methods, people and other resources that transforms some input (substrate materials, chemicals, machines, equipments, energy, manpower, etc.) into an output (deposited mass, compositions, structures, properties) that has one or more observable responses. Some of the process variables (x_1, x_2,.., x_p) are controllable (temperature, reducing agent, metal source, pH, bath load, etc.), i.e. they can be suitably changed to control the process

Figure 6.2 General model of a process/system

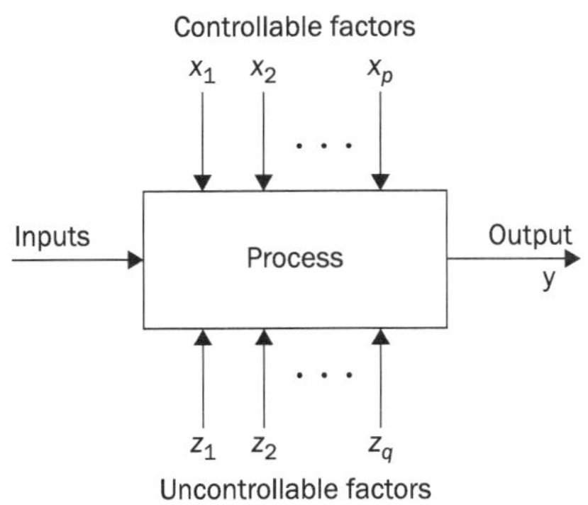

response, whereas other variables (z_1, z_2, . . ., z_n) are uncontrollable (errors in measuring instrument, impurities in substrates, human errors, etc.), i.e. they cannot be controlled in any way.

A designed experiment is a test or series of tests where pre-planned changes are made to the controllable variables of a process or system so that the reason for changes in the response can be observed and identified. Therefore, DOE refers to the systematic and scientific methods that are followed for planning experiments such that the experiments can be performed in the most efficient and economical way to get the required data that will result in valid and objective conclusions. This type of statistical approach to experimental design is required if we wish to draw a meaningful conclusion from the observed data.

DOE can be highly effective (Montgomery, 2001) when it is necessary to:

- Optimize product and process designs;
- Study the effects of multiple factors on the performance;
- Solve production problems by objectively laying out the experiments;
- Study the influence of the individual factors on the performance;
- Study the significance of each factor on the system response;
- Establish the tolerance levels of the factors;
- Allocate quality assurance resources based on the objective data;
- Indicate whether a supplier's part causes problems or not;
- Combine different factors in their proper settings to get the best results;

6.5.2 Taguchi method

The Taguchi technique (Ross, 1996; Taguchi, 1990) is a powerful tool for the design of high quality systems based on orthogonal array experiments that provide much reduced variance for the experiments with an optimum setting of process control parameters. This method achieves the integration of DOE with the parametric optimization of the process yielding the desired results. The orthogonal array (OA) requires a set of well balanced (minimum experimental runs) experiments. In this method, main parameters, which are assumed to have an influence on process results, are located at different rows in a designed OA. With such an array, completely randomized experiments can be conducted. Taguchi's method uses the statistical measure of performance called signal-to-noise ratios (*S/N*), which are logarithmic functions of desired output to serve as objective functions for optimization. The ratio depends on the quality characteristics of the product/process to be optimized. Three categories of *S/N* ratios are used: lower-the-better (LB), higher-the-better (HB) and nominally-the-best (NB). The parameter level combination that maximizes the appropriate *S/N* ratio is the optimal setting. For the case of minimization of friction or wear, the LB characteristic needs to be used.

6.5.3 Grey relational theory

Single response (individual friction and wear performance) optimization problems can be efficiently handled by the Taguchi method. But when both friction and wear are to be minimized simultaneously, the problem becomes a multiple response optimization. It is visibly different from that of a single performance optimization, as higher *S/N* ratio for one performance characteristic may correspond to a lower *S/N* ratio for another. Therefore, the overall evaluation of the *S/N* ratio is required for the optimization of multiple performance characteristics. Grey relational analysis is an efficient tool for this type of multi-response analysis. Grey relational analysis owes its origin to grey system theory which was first proposed by Deng (Deng, 1989). Any system in nature is not white (full of precise information), but on the other hand it is not black (completely lacking of information) either, and is mostly grey (a mixture of black and white). The incompleteness of information is the basic characteristic and serves as the fundamental starting point of the investigation of the grey system. Incomplete information follows from the limited availability of data, and the central problem of grey system theory is to seek only the intrinsic structure of the system given such limitation of data. The main objective of grey system theory is to supply information so that one can whiten the greyness. Grey relational analysis is based on the grey system theory, and compares and computes the dynamic causalities of the subsystems of a given system.

In grey relational analysis the first step is to perform the grey relational generation in which the results of the experiments are normalized in the range between zero and one. Then the second step is to calculate the grey relational coefficient from the normalized data to represent the correlation between the desired and actual experimental data. The overall grey relational grade is then computed by averaging the grey relational coefficient corresponding to each performance characteristic. Overall evaluation of the multiple performance characteristics is based on the calculated grey relational grade. As a result, optimization of the complicated multiple performance characteristics is converted into optimization of a single grey relational grade. The optimal level of the process parameters is the level with the highest grey relational grade. Furthermore, a statistical analysis of variance (ANOVA) (Montgomery, 2001) is performed to find which process parameters are statistically significant. With the grey relational analysis and ANOVA, the optimal combination of the process parameters can be predicted. Finally, a confirmation experiment is conducted to verify the optimal process parameters obtained from the analysis.

6.5.4 Deposition of Ni-P coating

Mild steel specimens are used as the substrate material for the deposition of the EN coating. The bath composition and operating conditions for EN coating are selected after several experiments and the proper ranges of the parameters are chosen accordingly. EN deposition is carried out using nickel chloride and nickel sulphate (concentration 30 g/l) as the source of nickel, sodium hypophosphite (concentration 10 g/l) as the reducing agent and sodium succinate (concentration 12 g/l) as the stabilizer. The concentration of stabilizer used in baths is kept fixed. The pH value of the bath is maintained at a fixed value of 4.5 by adding the required quantity of dilute hydrochloric acid. The cleaned samples are activated in palladium chloride at 55°C temperature and placed in the bath (volume 175 ml) for deposition for two hours at a constant temperature of 80°C. Deposition time is kept constant for each specimen so that the coating thickness remains approximately constant and the average coating thickness is found to be around 35 μm (Fig. 6.3). After the deposition, the samples are taken out of the electroless nickel bath and washed in distilled water. Then the samples are heat treated in a box furnace. Each specimen is coated and heat treated in air (annealed at 400°C for 1 hour) separately.

Figure 6.3 Cross-sectional view of Ni-P deposit

6.5.5 Design factors

Design factors are the controllable factors that are suitably varied in order to obtain the desired performance. There are a large number of factors that can be considered for controlling the tribological behavior of EN coatings. However, a review of the literature shows that the following three parameters are the most widespread among the researchers to control the tribological performance of EN coatings: load (L), speed (S) and time (T). These three factors are considered as main design factors along with their interactions in the present study. Table 6.3 shows the design factors along with their levels. The levels and ranges of the design factors are so selected considering the smaller thickness of the coatings (around 35 μm), otherwise through coating wear may occur. The present study is carried out to consider the tribological behavior of EN coatings as the performance characteristics. Thus, the response variables selected are coefficient of friction and wear depth.

6.5.6 Response variable

Response variable is the output of an experimental model. The evaluation of the tribological performance of EN coating

Table 6.3 **Design factors and their levels**

Design factors	Unit	Levels		
		1	2	3
Load (L)	N	50	75[a]	100
Speed (S)	RPM	60	70[a]	80
Time (T)	Min.	5	10[a]	15

a: initial testing condition

is carried out with friction and wear as response. Hence, coefficient of friction (COF) and wear depth (in micron) are used to represent both the responses respectively. The tribological testing parameters are optimized with the objective of minimizing both the response variables.

6.5.7 *Forming the design of experiments (DOE)*

DOE refers to the process of planning, designing and analyzing the experiment so that valid and objective information can be drawn effectively and with greater efficiency. In performing DOE, intentional changes are made to the controllable variables or factors in order to observe the corresponding changes in the output response. Hence, the maximum amount of conclusive information can be obtained in an organizing manner from the minimum amount of work, time, energy, money or other limited resources. The DOE using Taguchi approach can economically satisfy the needs of problem solving and product/process design optimization projects in the manufacturing industry. By learning and applying this technique, it is possible to significantly reduce the time required for experimental investigations.

As mentioned earlier, the Taguchi method uses an OA to reduce the number of experiments for determining the optimal process parameters. The OA allows one to compute the main and interaction effects via a minimum number of experimental trials (Ross, 1996). Several standard OAs have been tabulated by Taguchi. For the present investigation, L_{27} OA is chosen based on the number of factors considered, their levels and the desired interactions of the factors. The selected array requires the execution of 27 experiments. The factors (L, S, T) and their interactions (L×S, L×T, S×T) are assigned to their respective positions in the OA (as seen in Table 6.4) based on the Triangular Table for three-level OA (Roy, 1990). The cell values 1, 2 and 3 as seen in the array correspond to the lowest, medium and highest levels of the factors.

6.5.8 Friction and wear tests

Friction and wear characteristics of the EN coated specimens are studied under dry, non-lubricated conditions and at ambient temperature of about 25°C in a multi-tribotester apparatus (DUCOM, India) using a plate-on-roller configuration. A schematic diagram of the test rig is shown in Fig. 6.4. The EN coated stationary plates (size 20 mm / 20 mm / 8 mm) in the horizontal position are pressed against a rotating steel roller (diameter 50 mm / thickness 20 mm and composition conforming to EN8 specification) with hardness of 55 HRc, placed below the plate. In this method, the plate and the roller are arranged in such a way that the rotating roller serves as the counterface material while the stationary plate serves as the test specimen. A 1:5 ratio loading lever is used to apply normal load on the top specimen. The loading lever is pivoted near to the normal load sensor and carries counter weight at one end while at the other end a loading pan is suspended for placing the dead weights. The frictional force is measured by a frictional

Table 6.4 L_{27} orthogonal array with design factors and interactions assigned

	Column numbers												
Trial No.	1 (L)	2 (S)	3 L×S)	4 (L×S)	5 (T)	6 (L×T)	7 (L×T)	8 (S×T)	9 –	10 –	11 (S×T)	12 –	13 –
1	1	1	1	1	1	1	1	1	1	1	1	1	1
2	1	1	1	1	2	2	2	2	2	2	2	2	2
3	1	1	1	1	3	3	3	3	3	3	3	3	3
4	1	2	2	1	1	1	2	2	2	3	3	3	
5	1	2	2	2	2	2	2	3	3	3	1	1	1
6	1	2	2	2	3	3	3	1	1	1	2	2	2
7	1	3	3	3	1	1	1	3	3	3	2	2	2
8	1	3	3	3	2	2	2	1	1	1	3	3	3
9	1	3	3	3	3	3	3	2	2	2	1	1	1
10	2	1	2	3	1	2	3	1	2	3	1	2	3
11	2	1	2	3	2	3	1	2	3	1	2	3	1
12	2	1	2	3	3	1	2	3	1	2	3	1	2
13	2	2	3	1	1	2	3	2	3	1	3	1	2
14	2	2	3	1	2	3	1	3	1	2	1	2	3

(*Continued*)

Table 6.4 L_{27} orthogonal array with design factors and interactions assigned (*Continued*)

	Column numbers												
Trial No.	1 (L)	2 (S)	3 L×S)	4 (L×S)	5 (T)	6 (L×T)	7 (L×T)	8 (S×T)	9 –	10 –	11 (S×T)	12 –	13 –
15	2	2	3	1	3	1	2	1	2	3	2	3	1
16	2	3	1	2	1	2	3	3	1	2	2	3	1
17	2	3	1	2	2	3	1	1	2	3	3	1	2
18	2	3	1	2	3	1	2	2	3	1	1	2	3
19	3	1	3	2	1	3	2	1	3	2	1	3	2
20	3	1	3	2	2	1	3	2	1	3	2	1	3
21	3	1	3	2	3	2	1	3	2	1	3	2	1
22	3	2	1	3	1	3	2	2	1	3	3	2	1
23	3	2	1	3	2	1	3	3	2	1	1	3	2
24	3	2	1	3	3	2	1	1	3	2	2	1	3
25	3	3	2	1	1	3	2	3	2	1	2	1	3
26	3	3	2	1	2	1	3	1	3	2	3	2	1
27	3	3	2	1	3	2	1	2	1	3	1	3	2

Figure 6.4 Schematic diagram of experimental set-up

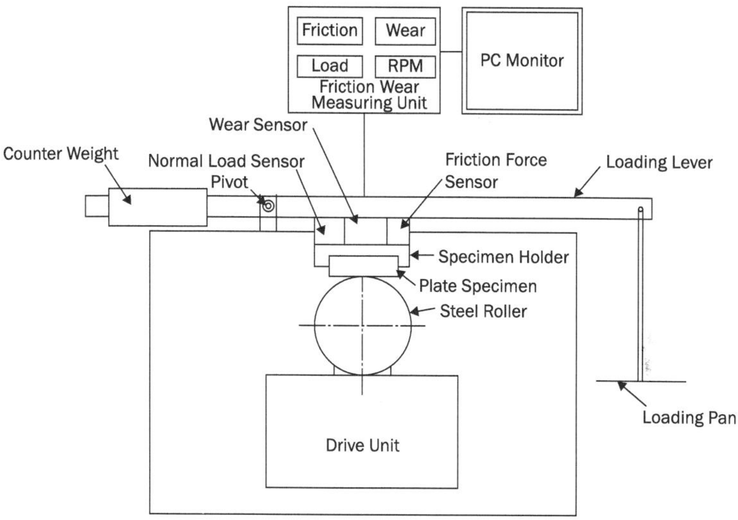

force sensor that uses a beam type load cell of capacity 1000 N.

Wear is measured in terms of displacement with the help of a linear voltage resistance transducer. The wear displacement sensor allows direct measurement of the loading lever's deflection, which corresponds to the wear of the specimen plate plus the wear of the counterface surface. The bottom roller being fully hardened undergoes negligible wear and thus the measured wear is essentially the wear of the coated plate specimen. It may be noted here that wear behavior is normally expressed as wear volume or weight loss, while in the present experimental set-up wear is measured in terms of displacement. Thus, to assess the accuracy of wear measurement, the displacement results for wear are compared with weight loss and show an almost linear relationship for the range of test parameters considered in the present study. The experiments are conducted with varying load (L), speed (S) and time (T) according to the combination of the levels of these design factors mentioned in the OA. The experimental results for friction and wear are given in Table 6.5.

Table 6.5 Experimental results for COF and wear depth

Exp. No.	Friction coefficient	Wear (μm)
1	0.257	13.57
2	0.264	15.65
3	0.276	17.22
4	0.228	11.21
5	0.247	14.19
6	0.256	16.24
7	0.172	11.16
8	0.179	12.97
9	0.211	16.06
10	0.114	14.15
11	0.149	18.31
12	0.163	23.58
13	0.095	13.54
14	0.099	16.56
15	0.109	23.16
16	0.077	13.38
17	0.090	15.79
18	0.094	21.75
19	0.072	23.07
20	0.073	25.78
21	0.076	26.39
22	0.055	22.33
23	0.059	23.46
24	0.064	26.30
25	0.042	21.91
26	0.044	22.19
27	0.049	24.46

6.5.9 Microstructural characterization

Surface morphology study of the EN coatings is carried out by scanning electron microscopy (SEM) (JEOL, JSM-6360) in order to analyze the microstructure of the deposited coatings before and after annealing to see the effect of heat treatment. SEM is also carried out after tribological testing to see the wear track patterns. Energy dispersive X-ray analysis (EDX) is carried out in conjunction with SEM to study the composition of the EN coatings in terms of the percentages of nickel and phosphorus in the coatings before and after heat treatment as well as after tribological testing. EDX is carried out in an energy dispersive X-ray analyzer (Inca, Oxford). An X-ray diffraction (XRD) analyzer (Rigaku, Miniflex) is used for identification of compounds in the EN coatings both before and after heat treatment.

Fig. 6.5 shows the SEM micrographs of the coating surfaces. The samples as deposited, as well as heat treated at 400°C, are studied by SEM. From the SEM micrographs of these surfaces it is seen that there are many globular particles on the surface of the substrate. The surface is optically smooth and of low porosity. No obvious surface damage is found. The surface of the Ni-P coatings appears to be dense. The XRD analysis (Fig. 6.6) indicates that the as-deposited Ni-P film is a mixture of nanocrystalline (uniformly spaced grain structure), microcrystalline phases. When Ni-P deposits are heat treated both their crystal structure and microstructure undergo modification. The microcrystalline deposits undergo a crystal growth process and such heat treatment results in a mixture of relatively coarse-grained structure.

The percentages of nickel and phosphorus of the EN deposits are determined using an EDX coupled to the SEM. It is seen that the as-deposited coating contains 88.6% Ni and 11.4% P while heat-treated (at 400°C) coatings have

Figure 6.5 SEM micrographs of the coating surfaces: (a) as deposited, (b) heat treated at 400°C

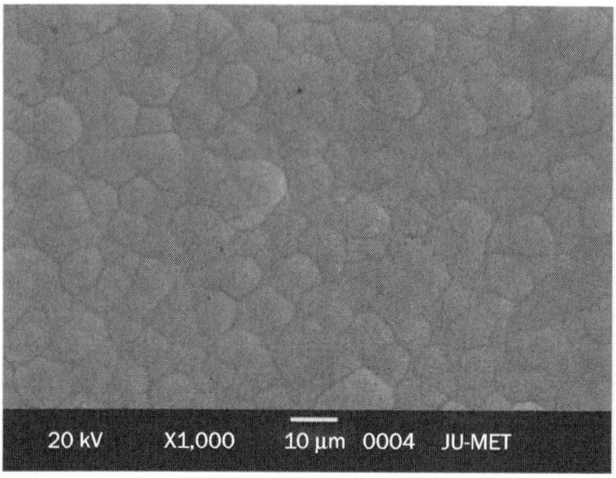

(a)

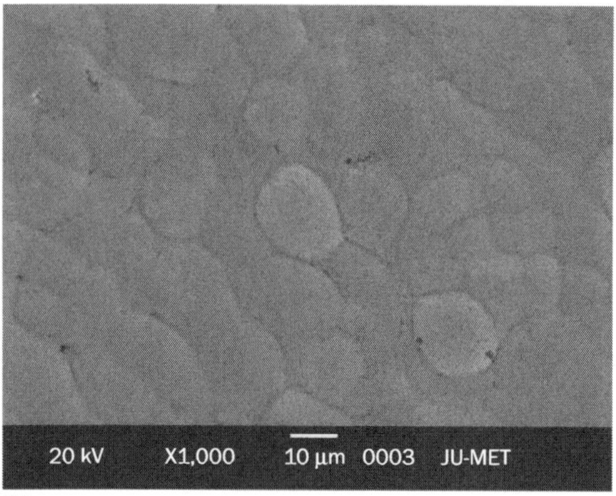

(b)

90.8% Ni and 9.2% P. The results show that the coatings contain more than 9 per cent of phosphorus leading to a nanocrystalline structure. The XRD patterns of Ni-P deposits in both as-plated and heat treated at 400°C are shown in Fig. 6.6. It is evident from Fig. 6.6(a) that Ni-P deposit is nanocrystalline in the as-plated condition. There is long range order giving rise to coherent scattering, although the domains are small in size. The mean grain size is of the order of 2 nm. Upon heat treatment at 400°C, Ni-P deposits crystallize and produce Ni_5P_2, Ni_2P and NiP_2 as major compound constituents, as seen in Fig. 6.6(b).

6.5.10 Analysis of signal to noise ratio

The primary goal of Taguchi method is to keep the variance in the output response very low, even in the presence of noise inputs. Hence, Taguchi suggests that the *S/N* ratio be used to convert the experimental results into a value for the evaluation characteristic in the optimum parameter analysis.

Figure 6.6 X-ray diffraction patterns of EN deposit: (a) as deposited, (b) heat treated at 400°C

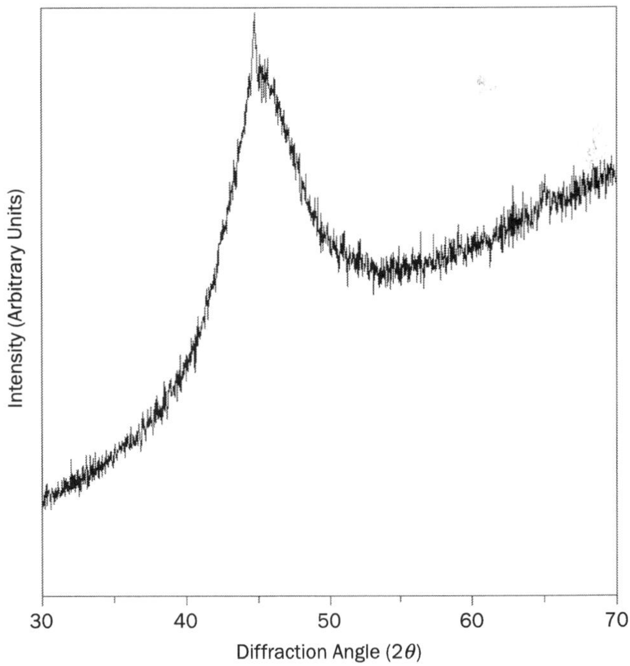

(a)

(Continued)

Figure 6.6 X-ray diffraction patterns of EN deposit (*Continued*)

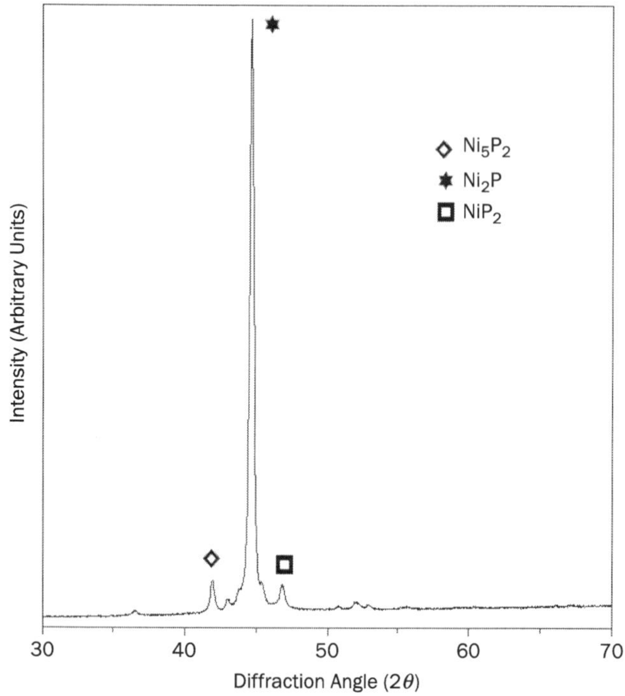

(b)

S/N ratio represents the ratio of sensitivity to variability. The higher the *S/N* ratio, the better is the quality of product. The idea is to maximize the *S/N* ratio, thereby minimizing the effect of random noise factors, which have a significant impact on the process performance. For the optimization of the individual tribological performances such as friction and wear, *S/N* ratio is carried out with COF and wear depth as the performances index. As both the parameters are desired to be minimized, *S/N* ratio is calculated with LB, the criterion which is given by Eq. [6.9]. In the case of multi-response optimization (combined friction and wear), *S/N* ratio analysis is carried out with the grey relational grade as the performance index. The *S/N* ratio for overall grey relational grade is calculated using the HB (higher-the-better) criterion given by Eq. [6.10].

$$S/N = -10\log\left(\frac{1}{n}\sum y^2\right) \qquad [6.9]$$

$$S/N = -10\log\left(\frac{1}{n}\sum \frac{1}{y^2}\right) \qquad [6.10]$$

where y is the observed data and n is the number of observations. The *S/N* values obtained from the given relations are used to construct various plots like the main effect plots and the interaction effects plots.

Main effects plots

The main effects plots give the optimal combination of testing parameters for the desired tribological performance such as minimum COF, wear, etc. The optimal combination of parameters for each of the performance studies is given in Table 6.6. As many researchers have already seen that friction coefficient decreases with increase in applied load (Staia et al., 1996), the optimized level of load (L3) being the highest is quite consistent with the earlier findings. Also it can be noted that all the performance optimizations have highest level of speed (S3) and lowest level of time (T1)

Table 6.6 **Optimized performance results**

Performance considered	Optimized condition
Friction	L3S3T1
Wear	L1S3T1
Friction and Wear	L3S3T1

as part of their optimal combinations. This particular combination of speed and time may help to achieve a particular tribological condition which is extracting enhanced tribological performance from the coating.

The main effects plots also give a rough idea about the relative significance of the parameters on the system response. In the main effects plots, if the line for a particular parameter is near horizontal, then the parameter has no significant effect. On the other hand, a parameter for which the line has the highest inclination will have the most significant effect. The main effects plots for each of the optimization problems are given in Fig. 6.7. From Fig. 6.7(a), it can be seen that load and time affect the friction performance to a greater extent compared to the time. Fig. 6.7(b) shows that load and time affect the wear performance greatly whereas speed has minimum effect. In the case of combined performance (Fig. 6.7(c)), almost all the testing parameters are found to be influential.

Interaction effects plots

The interaction effects plots again give a qualitative idea about the degree of interaction present between two parameters. Estimating an interaction means determining the non-parallelism of parameter effects. Thus, if the lines on the interaction plots are non-parallel, interactions occur, and if the lines cross, strong interactions occur between the parameters. For brevity, the interaction plots are not provided in this chapter.

6.5.11 Analysis of variance (ANOVA)

ANOVA is a statistical technique which can offer some important conclusions based on analysis of the experimental

Figure 6.7 Main effects plot for (a) friction performance, (b) wear performance, (c) combined performance

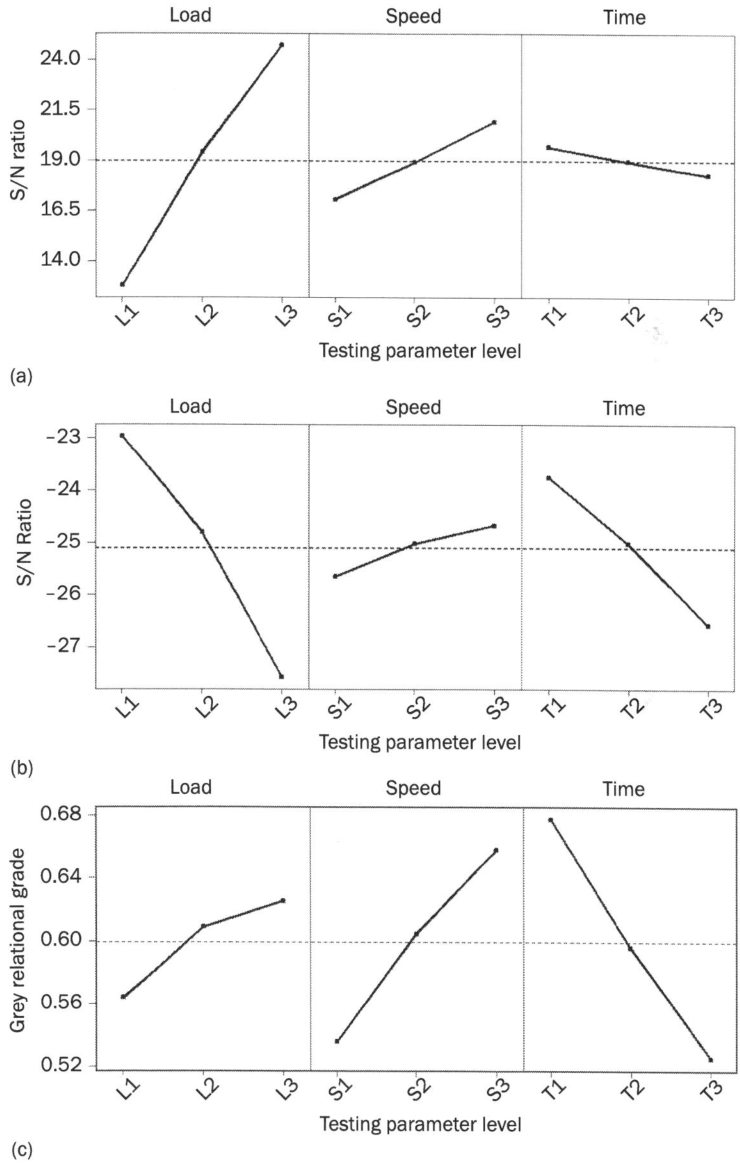

data. The method is very useful for revealing the level of significance of the influence of factor(s) or interaction of factors on a particular response. It separates the total variability of the response into contributions of each of the factors and the error. If some testing parameters do not considerably impact the tribological performance of EN coatings, they can be kept within a suitable range for the test and can be excluded in building future prediction and optimization models. ANOVA calculates the *F*-ratio, which is the ratio between the regression mean square and the mean square error. The *F*-ratio, also called the variance ratio, is the ratio of variance due to the effect of a factor and variance due to the error term. This ratio is used to measure the significance of the parameters under investigation with respect to the variance of all the terms included in the error term at the desired significance level α. If the calculated value of *F*-ratio is higher than the tabulated value of *F*-ratio, then the factor is significant at the desired level. In general, when the *F* value increases the significance of the parameter also increases. ANOVA also provides the percentage contribution of each parameter on the system response.

Friction performance

ANOVA provides a quantitative measure of the degree of influence of the design variables and their interactions on the process response. For the friction performance, ANOVA is performed with *S/N* ratio and it was found that all the three tribological testing parameters are significant at a confidence level of 99 per cent with parameter L (load) accounting for the highest contribution (88.53 per cent). Hence, friction performance of Ni-P coatings is very much dependent upon load and it should definitely be considered for building future friction models of Ni-P.

Wear performance

For the wear performance, ANOVA of the *S/N* ratio showed that all the three testing parameters are significant at the confidence level of 99 per cent. Parameters L (load) and time (T) with a contribution of 66 per cent and 24 per cent respectively, are the most significant factors influencing the wear performance of EN coating. The interaction between load and time (L×T) is the only significant interaction at 99 per cent confidence level and with a contribution of about 6 per cent. The importance of load in wear performance is quite expected as seen in earlier works, since more pressure on a surface in contact would definitely cause it to wear out more quickly.

Combined friction and wear performance

ANOVA for the combined tribological performance optimization was carried out using the overall grey relational grade of friction and wear response. It was again found that all the parameters significantly affect the performance at the confidence level of 99 per cent. Parameter T (time) is found to have the maximum contribution (47.27 per cent) while parameters S (speed) and L (load) have a contribution of 30.51 and 8.27 per cent. Similar to wear performance, interaction between load and time (L×T) is significant in this case and possesses a contribution of 10.74 per cent. The list of significant parameters and their contributions for all the performance studies is given in Table 6.7 for easy reference.

6.5.12 Confirmation tests

After the optimal level of test parameters has been identified, a confirmation test needs to be carried out in order to check the accuracy of analysis. Its purpose is to verify that the

Table 6.7 **Significant parameters from ANOVA**

Tribo performance	Influential parameters (99% confidence level)	Percentage contribution
Friction	L	88.53
	S	9.20
	T	1.23
Wear	L	66.00
	T	24.00
	S	3.00
	L×T	6.00
Friction and Wear	T	47.27
	S	30.51
	L	8.27
	L×T	10.74

optimum conditions suggested by the matrix experiment do indeed give the improvement projected. The verification experiment is performed by conducting a test with optimal settings of the factors and levels previously evaluated. The predicted value ($\hat{\eta}$) of the response at the optimum level is calculated with the Eq. [6.12].

$$\hat{\eta} = \eta_m + \sum_{i=1}^{o} \left(\bar{\eta}_i - \eta_m \right) \qquad [6.12]$$

where η_m is the total mean *S/N* ratio (for individual friction and wear performance) or total mean grey relational grade (for combined performance), $\bar{\eta}_i$ is the mean *S/N* ratio (for individual performance) or mean grey relational grade (for combined performance) at the optimal level, and o is the number of main design parameters that significantly affect the desired performance of EN coating.

Table 6.8 shows results of the confirmation tests for different performances. For friction performance, the

Table 6.8 **Results of confirmation tests**

	Initial parameter combination	Optimal parameter combination	
		Prediction	Experimentation
Setting level COF S/N ratio (dB)	L2S2T2 0.098 20.122	L3S3T1 27.371	L3S3T1 0.042 27.595
Setting level Wear (μm) S/N ratio (dB)	L2S2T2 16.563 –24.382	L1S3T1 –21.164	L1S3T1 11.163 –20.955
Setting level COF Wear (μm) Grey grade	L2S2T2 0.099 16.563 0.629	L3S3T1 0.762	L3S3T1 0.042 21.914 0.707

predicted *S/N* ratio is found to be in good agreement with the actual (experimental) *S/N* ratio at the optimal parameter combination. The improvement of *S/N* ratio from initial to optimal condition is 7.473 dB which implies COF is reduced by about 57 per cent. In case of the wear, performance improvement of *S/N* ratio from initial to optimal condition is 3.427 dB and the corresponding the reduction in wear is found to be around 32 per cent. Similarly, in case of the combined performance improvement of grey relational grade is found to be 0.078. This is about 13 per cent of the mean grey relational grade. Thus it can be seen that the optimization technique could be successfully applied in all the three cases and the corresponding improvement in the tribological performance is significant.

6.5.13 Wear mechanism

The SEM micrographs and EDX spectrum of the worn surface of coating are shown in Fig. 6.8 and Fig. 6.9

Figure 6.8 SEM micrograph of wear tested sample at: (a) 50 N load, (b) 75 N load, (c) 100 N load

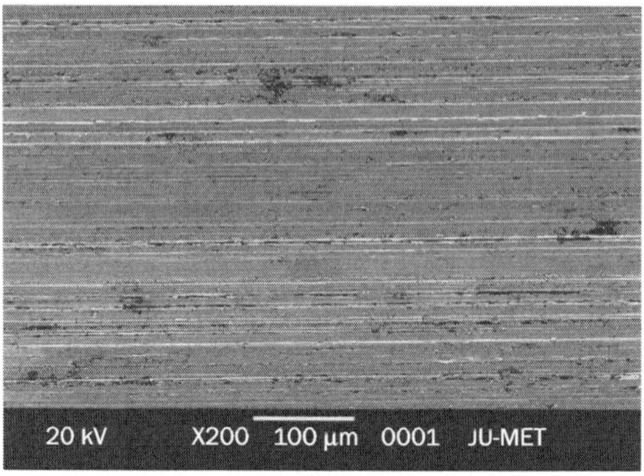

(a)

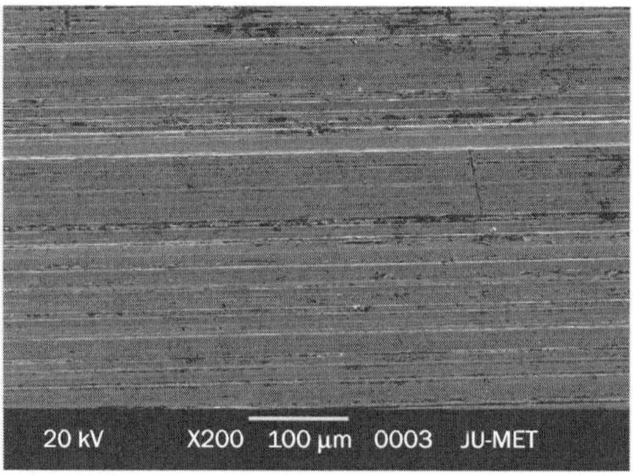

(b)

(c)

Figure 6.9 EDX pattern of tested sample

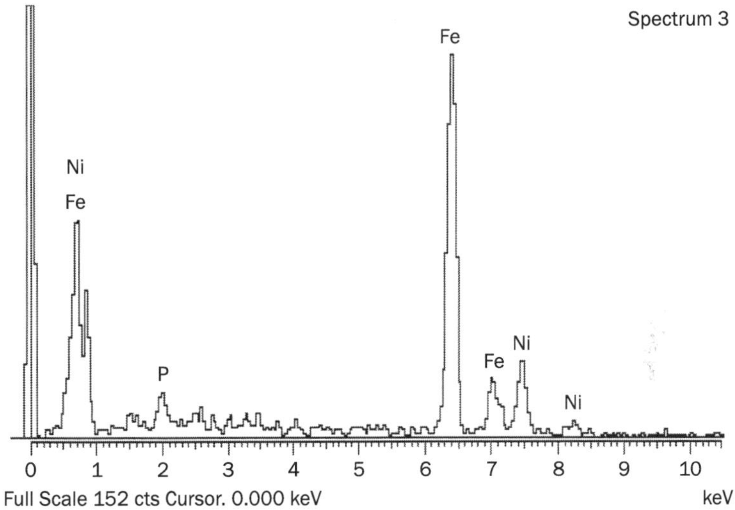

respectively. The SEM micrographs of coating tested at different loads exhibit a similar pattern. It is observed that the worn surface is mainly composed of longitudinal grooves and partial irregular pits along the sliding direction. The presence of grooves indicates the micro-cutting and micro-ploughing effect of the counterface while pits or prows are indicative of adhesive wear failure of the EN coating. The adhesive wear occurs under the experimental conditions used that induce a substantial attractive force between the mating surfaces leading to a high mutual solubility of nickel and iron. The EDX spectrum of the worn surface in Fig. 6.9 confirms the presence of iron. Thus, wear mechanism is found to be a combination of abrasive and adhesive wear.

6.6 Conclusions

In this chapter a detailed discussion of the tribological performance (friction and wear) of EN coatings is presented.

Optimization of the individual performances by varying three tribological testing parameters viz. load (L), speed (S) and time (T) is carried out with the help of Taguchi method. Optimization is also carried out for multiple performance characteristics (combined friction and wear) combining grey relational analysis with Taguchi method. The optimal combination of testing parameters is found to be L3S3T1, both for individual friction performance as well as for combined friction and wear performance. For individual wear performance, the optimal combination is found to be L1S3T1. From ANOVA it is seen that all the parameters have a significant impact on the performances at a confidence level of 99 per cent with load (L) being the highest contributor on the individual friction and wear performances and time (T) to the combined performance. The interaction between load and time (L×T) is the lone significant interaction in case of individual wear and the combined performance. The confirmation test shows a reduction of about 57 per cent and 32 per cent in COF and wear values respectively for the individual performance studies. For combined performance, the improvement of grey relational grade is found to be 13 per cent compared to the mean grey grade. SEM of the coating surface reveals a globular structure. The surface also appears to be of low porosity, smooth and with no surface damage. XRD analysis shows that as-deposited Ni-P film is a mixture of nanocrystalline (uniformly spaced grain structure) and microcrystalline phases which turn crystalline with heat treatment giving rise to a relatively coarse-grained structure. The EDX analysis shows that the films contain more than 9 per cent of phosphorus resulting in a nanocrystalline structure. The wear mechanism in Ni-P coatings is found to be a combination of abrasive and adhesive wear.

6.7 References

Agarwala, R.C. and Agarwala, V. (2003), 'Electroless alloy/composite coatings: A review', *Sadhana – Academy Proceedings in Engineering Sciences*, **28**(3–4): 475–93.

Alirezaei, S., Monirvaghefi, S.M., Salehi, M. and Saatchi, A. (2007), 'Wear behavior of Ni-P and Ni-P–Al_2O_3 electroless coatings', *Wear*, **262**: 978–85.

Alirezaei, S., Monirvaghefi, S.M., Salehi, M., Saatchi, A. and Kargosha, M. (2005), 'Effect of alumina content on wear behaviour of Ni-P-Al_2O_3 electroless composite coatings', *Surface Engineering*, **21**(1): 60–6.

Allen, R.M. and Vandersande, J.B. (1982), 'The structure of electroless Ni-P films as a function of composition', *Scripta Metallurgica et Materiala*, **16**(10): 1161–4.

Anik, M., Körpe, E. and Şen, E. (2008), 'Effect of coating bath composition on the properties of electroless nickel-boron films', *Surface and Coatings Technology*, **202**(9): 1718–27.

Apachitei, I., Duszczyk, J., Katgerman, L. and Overkamp, P.J.B. (1998), 'Electroless Ni-P composite coatings: The effect of heat treatment on the microhardness of substrate and coating', *Scripta Mater*, **38**(9): 1347–53.

Araghi, A. and Paydar, M.H. (2010), 'Electroless deposition of Ni-P-B_4C composite coating on AZ91D magnesium alloy and investigation on its wear and corrosion resistance', *Materials and Design*, **31**(6): 3095–9.

Balaraju, J.N., Sankara Narayanan, T.S.N. and Seshadri, S.K. (2003), 'Electroless Ni-P composite coatings', *Journal of Applied Electrochemistry*, **33**(9): 807–16.

Baudrand, D. and Bengston, J. (1995), 'Electroless plating processes. Developing technologies for electroless nickel, palladium, and gold', *Metal Finishing*, **93**(9): 55–7.

Berkh, O., Eskin, S., and Zahavi, J. (1996), 'Properties of electrodeposited Ni-P-SiC composite coatings', *Met. Finish.*, **94**: 35–40.

Brenner, A. and Riddell, G. E. (1950), *Nickel Plating by Chemical Reduction*. US2532282 (patent).

Das, C.M., Limaye, P.K., Grover, A.K. and Suri, A.K. (2007), 'Preparation and characterization of silicon nitride codeposited electroless nickel composite coatings', *Journal of Alloys and Compounds*, **436**(1–2): 328–34.

Delaunois, F. and Lienard, P. (2002), 'Heat treatments for electroless nickel-boron plating on aluminium alloys', *Surface and Coatings Technology*, **160**(2–3): 239–48.

Delaunois, F., Petitjean, J.P., Lienard, P. and Jacob-Duliere, M. (2000), 'Autocatalytic electroless nickel-boron plating on light alloys', *Surface and Coatings Technology*, **124**(2–3): 201–9.

Deng, J. (1989), 'Introduction to grey system', *Journal of Grey System*, **1**(1): 1–24

Dong, D., Chen, X.H., Xiao, W.T., Yang, G.B. and Zhang, P.Y. (2009), Preparation and properties of electroless Ni-P-SiO_2 composite coatings', *Applied Surface Science*, **255**(15): 7051–5.

Ebrahimian-Hosseinabadi, M., Azari-Dorcheh, K. and Vaghefi, S.M.M. (2006), 'Wear behavior of electroless Ni-P-B_4C composite coatings', *Wear*, **260**(1–2): 123–7.

Fisher, R.A. (1951), *Design of Experiments*, Edinburgh: Oliver and Boyd.

Gay, P.A., Limat, J.M., Steinmann, P.A. and Pagetti, J. (2007), 'Characterisation and mechanical properties of electroless NiP-ZrO_2 coatings', *Surface and Coatings Technology*, **202**: 1167–71.

Hamid, Z.A., El Badry, S.A. and Aal, A.A. (2007), 'Electroless deposition and characterization of Ni-P-WC composite alloys', *Surface and Coatings Technology*, **201**(12): 5948–53.

Holmberg, K. and Matthews, A. (1998), *Coatings Tribology*, Amsterdam: Elsevier.

Hu, X., Jiang, P., Wan, J., Xu, Y. and Sun, X. (2009), 'Study of corrosion and friction reduction of electroless Ni-P coating with molybdenum disulfide nanoparticles', *Journal of Coatings Technology Research*, **6**(2): 275–81.

Huang, Y.S., Zeng, X.T., Annergren, I. and Liu, F.M. (2003), 'Development of electroless NiP-PTFE-SiC composite coating', *Surface and Coatings Technology*, **167**(2–3): 207–11.

Hur, K.H., Jeong, J.H. and Lee, D.N. (1990), 'Microstructures and crystallization of electroless Ni-P deposits', *J Mater Sci*, **25**(5): 2573–84.

Krishnaveni, K., Sankara Narayanan, T.S.N. and Seshadri, S.K. (2005), 'Electroless Ni-B coatings: Preparation and evaluation of hardness and wear resistance', *Surface and Coatings Technology*, **190**(1): 115–21.

Li, Z., Chen, Z., Liu, S., Zheng, F. and Dai, A. (2008), 'Corrosion and wear properties of electroless Ni-P plating layer on AZ91D magnesium alloy', *Transactions of Nonferrous Metals Society of China* (English Edition), **18**(4): 819–24.

Madej, M., Ozimina, D. and Piwoński, I. (2006), 'The influence of tribochemical reactions of antiwear additives on heterogeneous surface layers in boundary lubrication', *Tribology Letters*, **22**(2): 135–41.

Mallory, G.O. and Hadju, J.B. (1991), *Electroless Plating: Fundamentals and Applications*, Orlando, FL: AESF.

Montgomery, D.C. (2001), *Design and Analysis of Experiments*, New York: Wiley.

Novák, M., Vojtěch, D. and Vítů, T. (2010), 'Influence of heat treatment on tribological properties of electroless

Ni-P and Ni-P-Al_2O_3 coatings on Al-Si casting alloy', *Applied Surface Science*, **256**(9): 2956–60.

Palaniappa, M. and Seshadri, S.K. (2008), 'Friction and wear behavior of electroless Ni-P and Ni-W-P alloy coatings', *Wear*, **265**(5–6): 735–40.

Quercia, G., Grigoresu, I., Contreras, H., Rauso, C.D. and Gutierrez-Campos, D. (2001), 'Friction and wear behavior of several hard materials', *International Journal of Refractory Metals and Hard Materials*, **19**: 359–69.

Ramalho, A. and Miranda, J.C. (2005), 'Friction and wear of electroless NiP and NiP + PTFE coatings', *Wear*, **259**(7–12): 828–34.

Riedel, W. (1991), *Electroless Nickel Plating*, Stevenage: Finishing Publications Ltd.

Ross, P.J. (1996), *Taguchi Techniques for Quality Engineering*, 2nd edn, New York: McGraw-Hill.

Roy, R.K. (1990), *A Primer on the Taguchi method,* Dearborn: Society of Manufacturing Engineers.

Sahoo, P. and Das, S.K. (2011), 'Tribology of electroless nickel coatings – A review', *Materials and Design*, **32**: 1760–75.

Sarret, M., Müller, C. and Amell, A. (2006), 'Electroless NiP micro- and nano-composite coatings', *Surface and Coatings Technology*, **201**(1–2): 389–95.

Srinivasan, K.N. and John, S. (2005), 'Studies on electroless nickel-PTFE composite coatings', *Surface Engineering*, **21**(2): 156–60.

Staia, M.H., Castillo, E.J., Puchi, E.S., Lewis, B. and Hintermann, H.E. (1996), 'Wear performance and mechanism of electroless Ni-P coating', *Surface and Coatings Technology*, **86–7**(2): 598–602.

Staia, M.H., Enriquez, C. and Puchi, E.S. (1997), 'Influence of the heat treatment on the abrasive wear resistance of electroless Ni-P', *Surface and Coatings Technology*, **94–5**: 543–8.

Straffelini, G., Colombo, D. and Molinari, A. (1999), 'Surface durability of electroless Ni-P composite deposits', *Wear*, **236**(1–2): 179–88.

Taguchi, G. (1990), *Introduction to Quality Engineering*, Tokyo: Asian Productivity Organization.

Taheri, R., Oguocha, I.N.A. and Yannacopoulos, S. (2001), 'The tribological characteristics of electroless NiP coatings', *Wear*, **249**(5–6): 389–96.

Tsujikawa, M., Azuma, D., Hino, M., Kimura, H. and Inoue, A. (2005), 'Friction and wear behavior of laser irradiated amorphous metal', *Journal of Metastable and Nanocrystalline Materials*, **24–5**: 375–8.

Wu, Y., Liu, H., Shen, B., Liu, L. and Hu, W. (2006a), 'The friction and wear of electroless Ni-P matrix with PTFE and/or SiC particles composite', *Tribology International*, **39**(6): 553–9.

Wu, Y., Shen, B., Liu, L., and Hu, W. (2006b), 'The tribological behaviour of electroless Ni-P-Gr-SiC composite', *Wear*, **261**(2): 201–7.

Ying, H.G., Yan, M., Ma, T.Y., Wu, J.M. and Yu, L.Q. (2007), 'Effects of NH_4F on the deposition rate and buffering capability of electroless Ni-P plating solution', *Surface and Coatings Technology*, **202**(2): 217–21.

Zhao, Q., Liu, Y., Müller-Steinhagen, H. and Liu, G. (2002), 'Graded Ni-P-PTFE coatings and their potential applications', *Surface and Coatings Technology*, **155**(2–3): 279–84.

Zou, T.Z., Tu, J.P., Zhang, S.C., Chen, L.M., Wang, Q. et al. (2006), 'Friction and wear properties of EN- (IF-MoS_2) composite coatings in humid air and vacuum', *Materials Science and Engineering A*, **426**(1–2): 162–8.

7

Tribological response of materials during sliding against various surface textures

P. Menezes, Kishore and S. Kailas, Indian Institute of Science, India, and M. Lovell, University of Wisconsin–Milwaukee, USA

Abstract: Surface texture plays an important role during sliding. In this investigation, the tribological response of materials with varying hardness and crystal structure during sliding against a relatively harder material of different surface textures was studied. It was found that the coefficient of friction is controlled by the surface texture of the harder mating surfaces. Further, two kinds of frictional response, namely steady-state and stick–slip, were observed during sliding. In general, the response was dependent on material pair, normal load, lubrication and surface texture of the harder material. The results can be applied to control the friction between the die and sheet metal in metal forming process.

Key words: friction, stick–slip, surface texture, surface roughness, transfer layer.

7.1 Introduction

7.1.1 Friction

Friction is the resistance to relative motion of two bodies that are in contact. More specifically, the friction is caused by forces acting at the interface between the surfaces of contacting bodies. The magnitude of these forces is related to the surface properties of the two contacting surfaces and the relative compliance of the materials in contact (Suh, 1986). The friction properties of pairs of materials are traditionally represented by the coefficient of friction, μ, defined as:

$$\mu = \frac{F}{N} \tag{7.1}$$

where F is the friction force and N the normal contact force.

The important factors that control friction are surface texture, normal load, sliding speed, environmental conditions such as temperature and lubricants, and material properties (Suh, 1986). Considerable work has been done to study the effect of these parameters on coefficient of friction using different experimental methods (Kayaba, 1962; Buckley, 1967; Lim et al., 1989; Dwivedi et al., 2004; Mokhtar et al., 1979, 1980; Menezes et al., 2006a, 2007, 2008a, 2009a, 2010, 2011; Lovell et al., 2010). Among many important factors which affect friction, the role of three parameters, namely surface texture, hardness and the crystal structure, are discussed in the present study.

7.1.2 Surface texture

Surface texture is the local deviation of a surface from a perfectly flat plane. The measure of the surface texture is generally determined in terms of its roughness, waviness and form (Bhushan, 2002). The influence of the surface texture on friction has not been well established. Early research in this direction mostly dealt with aspects like real area of contact and lubricant trapping. For example, in the theoretical work, using the plane–strain theory of plasticity for rigid-non-hardening materials, Kudo (1965) proposed a methodology to estimate the effect of trapped lubricant in microscopic surface pockets of a work material by the tool surface on the apparent coefficient of friction. It was also pointed out that the effect was more pronounced for lubricant having a higher bulk modulus. Nellemann et al. (1977) subsequently investigated the effect of different surface texture geometries by varying asperity angles and concluded that normal pressure and bulk modulus have great influence on the real area of contact, whereas the asperity slope and friction factor are of minor importance. Thus, these authors tried to present the concept of surface texturing. There are many ways to produce surface textures, including grinding, polishing, cutting, lapping, blasting, electric discharge marching (EDM), honing, milling, reaming, lithography and laser treatment. Surface textures are usually characterized in terms of roughness parameters using a profilometer.

7.1.3 Surface texture and friction in metal forming

Friction is a very important factor in metal forming processes. The coefficient of friction, if controlled properly, could generate the required stresses to deform the metal to the required shape. It could also lead to failure of the work-piece if not controlled properly. Sheet metal forming processes are increasingly focused on tribological parameters where the surface texture shows greater potential for process optimization. The specific contact condition for sheet metal forming is that the tool is continuously in contact with the work material. Efforts were made to study the effects of surface texture on friction during metal forming. Wagner (2001) studied the influence of tool surfaces, coating on tool surfaces, sheet metal surfaces, coating on sheet metal surfaces and the influence of the lubrication on the frictional behavior. Wagner (2001) summarized that the tribological conditions at the contact zones between the sheet surface and the tool surface play an important role in determining the limits of the forming process. It was concluded that the friction in the various tribologically relevant contact zones affects the flow of the material in the tool and hence it was used deliberately to control the drawing process.

7.1.4 Effect of surface texture of softer material on friction in metal forming

Several authors have reported on the effect of surface texture of soft material on the friction during metal forming. Rasp and Wichern (2002) studied the effect of surface texture on frictional resistance using different kinds of surfaces. In their experiment the five different specimen surfaces used were: as-received, etched, coarse ground perpendicular to test direction, coarse ground parallel to test direction, and polished. They found that the arithmetic roughness value (R_a) and lubrication regime have greater influence than the directionality of the surface lay. Bello and Walton (1987) studied the interaction of surface roughness and lubrication at the tool–metal interface in sliding contact. In their experiment, strips of commercial pure aluminum were pulled through steel dies designed to give partial simulation of the conditions that exist in the flange and die radius profile regions of the deep drawing process. They found that the conventional surface roughness parameters do not provide a satisfactory functional characterization of the surface in context of the friction developed in sliding contact. The influence of surface topography of the sheet material on the frictional characteristics of 3104 Al alloy sheet were investigated by Saha et al. (1996) by stretching a strip around a cylindrical pin. They found that friction increased with the strain occurring during the contact, which supports the model relating friction to flattening of strip asperities and real area of contact. They also found that the coefficient of friction depends on the rolling direction of the strip rather than surface topography. It is understood from the above literature that the surface texture of deformable materials cannot explain true friction values during sliding, and it is therefore important to have knowledge about the surface texture of harder materials on the coefficient of friction during sliding.

7.1.5 Effect of surface texture of harder material on friction in metal forming

Considerable amounts of work have been carried out to study the effect of surface texture of a harder material on the coefficient of friction during metal forming. Lakshmipathy and Sagar (1992) studied the influence of die grinding mark orientation on friction in open die forging under lubricated conditions. They used commercial pure aluminum as the work-piece material and H11 steel as the die material. Two sets of dies, one with unidirectional grinding marks and one with criss-cross grinding marks, were used. It was found that, for the same percentage of deformation, the dies with the criss-cross ground pattern required lower forging loads when compared with the die with a unidirectionally ground pattern. The friction factor was also lower during the forging process when the die with the criss-cross surface pattern was used. Lakshmipathy and Sagar (1992) concluded that the lubrication breakdown tendency is greater when pressing is performed with a unidirectionally ground die rather than with a criss-cross ground die. Määttä et al. (2001) studied the friction and adhesion of a stainless steel strip against different tool steels and reported that the composition of the tool steel does not have a marked effect on the friction between the tool and the work-piece. However, the surface topography of the tool had a marked effect; for example, polishing of the tool surface to reduce the surface roughness reduces the friction between the tool and the work-piece. Costa and Hutchings (2009) investigated the influence of surface texture on friction during metal forming processes. They concluded that the friction was strongly influenced by the relative orientation between the grooves generated on the die surfaces and the drawing direction. Hu and Dean (2000) studied the relation between friction behavior and surface topography using a ring upsetting test. The tests were carried out either using a liquid lubricant or under clean dry conditions. Two types of work-piece surfaces, random and directional, were prepared by either shot-blasting or EDM or turning to different levels of surface finish. It was found that for random surfaces, smoother surfaces could retain more lubricant and decrease the friction resistance. Malayappan and Narayanasamy (2004) studied the bulging effect of aluminum solid cylinders by varying the frictional conditions at the flat die surfaces. Different machining processes like grinding, milling, electro-spark machining and lathe turning with emery finish were used on the flat dies to vary the frictional conditions. It was concluded that the barreling depends on friction and ultimately the surface finish.

Attempts have also been made to simulate the tribological conditions that are encountered in metal forming operations by means of simple laboratory sliding tests. Staph et al. (1973) studied the effect of surface texture and surface roughness on scuffing using a 'caterpillar disc tester'. The authors used steel discs of varying roughness and texture (honed, circumferentially ground with low and high roughness, and cross-ground) and concluded that both surface texture and surface roughness affect frictional behavior. Koura (1980) studied the effect of surface texture on friction mechanism using a universal testing machine. Steel specimens were prepared to various degrees of roughness by grinding, lapping and polishing. The results showed that the behavior of surfaces and friction during sliding depends on the degree of roughness. Wakuda et al. (2003) studied the frictional properties of silicon nitride ceramic surfaces in which dimple patterns were machined with different size, density and geometry against hardened steel. They found that surfaces with dimples show reductions in friction coefficient when compared to lapped smooth surfaces. Wakuda et al. (2003) concluded that the tribological characteristics depended greatly on the size and density of the micro-dimples rather than the shape of the micro-dimples. Menezes et al. (2010) investigated the influence of directionality of surface grinding marks on friction under both dry and lubricated conditions. The authors concluded that the coefficient of friction significantly depended on the directionality of grinding marks on the harder steel surfaces.

7.1.6 Formation of transfer layer

When softer materials slide against harder materials, a transfer layer of the softer material forms on the harder counter-surface. Attempts have been made to study the formation of a transfer layer during sliding. Schedin (1994) conducted experiments using U-bending and strip drawing tests to study the formation of a transfer layer during forming processes. His experimental conditions resemble the contact conditions in sheet metal forming, where a hard and smooth tool surface will make repeated contact with a soft and rough sheet surface. He concluded that it is impossible to completely avoid metal transfer in the sheet metal forming operation, but that the growth of the transfer layer could be controlled by carefully designing the surface topography and using lubricants or coatings. Don et al. (1983) studied the friction and wear of Cu-Be and a dispersion hardened copper system and concluded that transfer material plays a significant role in both friction and wear for these systems. The effect of surface topography of metal disk on the film formation by transfer from plastic pads was investigated by Hayashi et al. (1999) using pad-on-disk type apparatus. The authors concluded that the roughness valleys across the sliding direction and the small pits on the disk surface accelerated the film formation of pad material on the disk.

7.1.7 Different frictional responses

In general, two kinds of frictional response, namely steady-state and stick–slip, can be observed during sliding. During steady-state, the frictional force remains constant with sliding distance or time. However, during stick–slip, the frictional force does not remain constant, but rather oscillates significantly as a function of sliding distance or time. During the stick phase, the friction force builds to a critical value. Once a critical force has been attained (to overcome the static friction), slip occurs at the interface and energy is released so that the frictional force decreases. This stick–slip phenomenon can occur if the coefficient of static friction is much greater than the coefficient of kinetic friction. Bowden and Tabor (1950) suggested that static friction is greater than kinetic friction due to molecular bonding between the surfaces. Bouissou et al. (1998) studied the influence of normal load, slip rate and roughness during sliding of self-mated polymethylmethacrylate (PMMA) under dry conditions. They observed a steady-state regime at low normal pressures and a stick–slip regime at high normal pressures for all slip rates and grades of roughness. For intermediate normal pressures, the transition between the two regimes was controlled by roughness and slip rates. Bouissou et al. (1998) concluded that normal pressure is the main parameter influencing the transition between stable sliding and stick–slip motion. Hwang and Gahr (2003) studied the static and kinetic friction for different pairs of bearing steel 100Cr6 and a commercial alumina under unlubricated and oil lubricated conditions as a function of normal loads and surface finish (ground and polished). Hwang and Gahr (2003) observed that both static and kinetic coefficient of friction were slightly lower under lubricated conditions than under unlubricated conditions. The value of static and kinetic coefficient of friction increases with increasing normal load for the ground surfaces. At a given value of normal load, the static coefficient of friction increases with increasing surface roughness. However, no effect of normal load was observed on pairs with polished surfaces, but the static coefficient of friction decreased slightly with increasing normal load on pairs with polished plates. Hwang and Gahr (2003) concluded that the stick–slip phenomena occurred with both unlubricated and lubricated pairs under high normal loads depending on the surface finish.

In the present investigation, the tribological response of different materials (Pb, Mg and Fe) with varying hardness and crystal structures during sliding against steel plates of different texture and roughness was studied. It is important to note that only the surface texture of a harder counter-surface was considered. The effect of these materials' responses on the coefficient of friction and formation of a transfer layer during sliding against different surface textures is addressed.

7.2 Experimental details

7.2.1 Surface textures

Three types of surface texture, namely (a) unidirectional, (b) 8-ground and (c) random, were attained on the EN8 steel plates. Unidirectional and 8-ground textures with varying roughness were produced on the steel plate by grinding the steel plates against different grits of emery papers (220, 400, 600, 800 or 1000 grit size). These different grits were used to vary the surface roughness of the steel plates. For the unidirectional case, care was taken so that the grinding marks were unidirectional in nature. The 8-ground surface was generated by moving the steel plate on emery papers along a path in a figure-of-eight motion for 500 cycles. The random textures were generated on the steel plates by moving the steel plate against the pad of disc polishing machine. To vary the surface roughness, five kinds of abrasive media (in slurry form) such as 220 grit SiC powder, 600 grit SiC powder, 1000 grit SiC powder, Al_2O_3 powder (0.017 μm), or diamond paste (1–3 μm) were used. Fig. 7.1 (a), (b) and (c) show the surface profiles of steel plates generated by unidirectional grinding, 8-ground and random grinding, respectively.

Figure 7.1 Surface profile of steel plates that are (a) unidirectionally ground (R_a = 0.29 μm), (b) 8-ground (R_a = 0.30 μm) and (c) randomly polished (R_a = 0.30 μm)

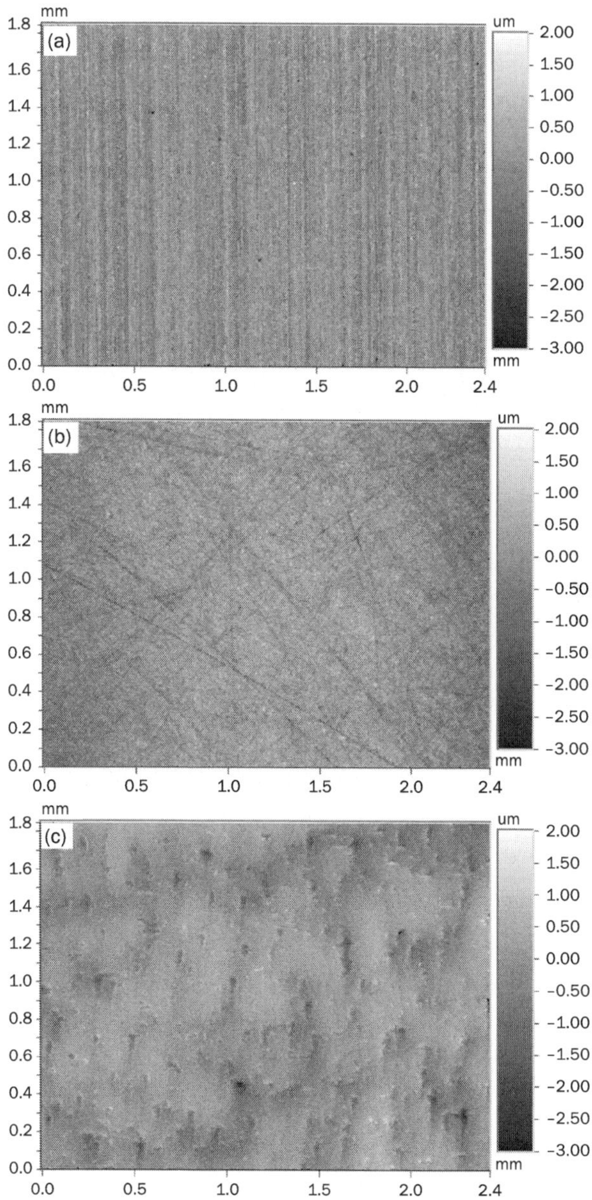

7.2.2 Materials

Three pin materials and a single plate material were used in this study and the details are presented in Table 7.1. The pins were 10 mm long, 3 mm in diameter with a tip radius of 1.5 mm. The dimensions of the counterpart steel plates were 28 mm × 20 mm × 10 mm (thickness). The pins were first machined and then electro-polished to remove any work-hardened layers that might have formed during the machining. Hardness measurements of the pins and steel plate were made at room temperature using a Vickers microhardness tester with 100 gm load and 10-second dwell time. Average hardness values, obtained from five indentations, were also presented in Table 7.1. Before each experiment, the pins and steel plates were first thoroughly cleaned with an aqueous soap solution and then with acetone in an ultrasonic cleaner.

7.2.3 Apparatus

Experiments were conducted using an inclined pin-on-plate sliding apparatus, the schematic view of which is shown in Fig. 7.2. The details of the pin-on-plate apparatus are presented elsewhere (Menezes et al., 2009b). The pin-on-plate sliding tester is extensively used for characterizing the tribological properties such as friction and/or wear. The technique involves (a) measuring both normal and tangential

Table 7.1 Details of the materials used for the study

	Material	Purity	Crystal Structure	Hardness (VHN)
Pins	Pb	99.98 wt. %	FCC	$8HV_{0.1}$
	Mg	99.98 wt. %	HCP	$55HV_{0.1}$
	Fe	99.9 wt. %	BCC	$148HV_{0.1}$
Plate	EN8	–	–	$208HV_{0.1}$

Figure 7.2 Schematic diagram of pin-on-plate with inclined steel plate

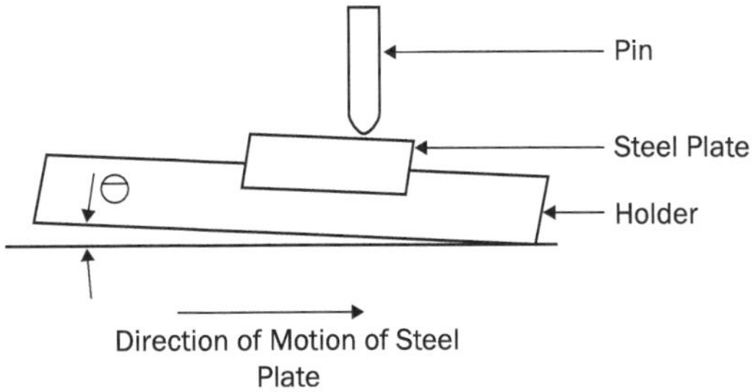

forces, and/or (b) generating wear tracks on hard plates using soft pins. In this study, the experiments were conducted using an inclined (i.e., progressively increasing load) sliding mode. Hence, the effect of load on the coefficient of friction can be readily determined in a single experiment. To perform the experiments, the steel plate was fixed horizontally in the vice of the pin-on-plate sliding tester and then the vice setup was tilted so that surface of the plate made an angle of 1° ± 0.1° with respect to horizontal base. The advantage of 1° inclination of the plate was that from a single experiment, the effect of normal load (up to the test limit of 120 N) on the coefficient of friction and the formation of transfer layer could be studied. The pins were slid at a speed of 2 mm/s against the prepared steel plate starting from the lower end to the higher end of the inclined surface for a sliding length of 10 mm. The waviness of the plate was not considered; this measured to be more than an order of magnitude lower than the plate angle. The normal and tangential forces were continuously acquired using a computer with a data acquisition system. The coefficient of friction was calculated using the formula given by Eq. [7.2].

$$\mu = \frac{F_T \cos\theta - F_N \sin\theta}{F_T \sin\theta + F_N \cos\theta} \qquad [7.2]$$

where 'θ' is the angle of inclination of the steel plate, F_T is the recorded traction force and F_N is the recorded normal force at any instance.

The pins were slid both in perpendicular and parallel directions to the unidirectional grinding marks on the plate. Thus, four sets of experimental conditions were used for a particular pin material. Experiments were conducted under both dry and lubricated conditions on each plate in an ambient environment. The dry tests were conducted first to avoid any additional cleaning of the steel plates and to exclude variations in roughness of the steel plates. After the dry tests, the pin was removed and a new pin from the same batch was mounted on the vertical slide to perform lubricated tests. For the lubricated tests, a drop (0.05 ml) of commercially available engine oil lubricant (Shell 2T oil) was applied on the surface of the same steel plate. The viscosity of lubricant oil was found to be 40 cSt at 40°C and had the extreme pressure additive, zinc dialkyl dithiophosphate (ZDDP). The presence of ZDDP was confirmed using the Fourier transform infrared spectroscopy technique. The profiles and surface roughness parameters of the steel plates were measured in the direction of the sliding on the bare surface away from the wear tracks using an optical profilometer. After the tests, the steel plates were observed using a scanning electron microscope (SEM) to reveal the morphology of the transfer layer.

7.3 Results

7.3.1 Variation of coefficient of friction with surface texture

Experiments were conducted by sliding pure Pb, pure Mg and pure Fe pins against steel plates of various surface

Figure 7.3 Variation of forces and coefficient of friction with sliding distance for pure Pb

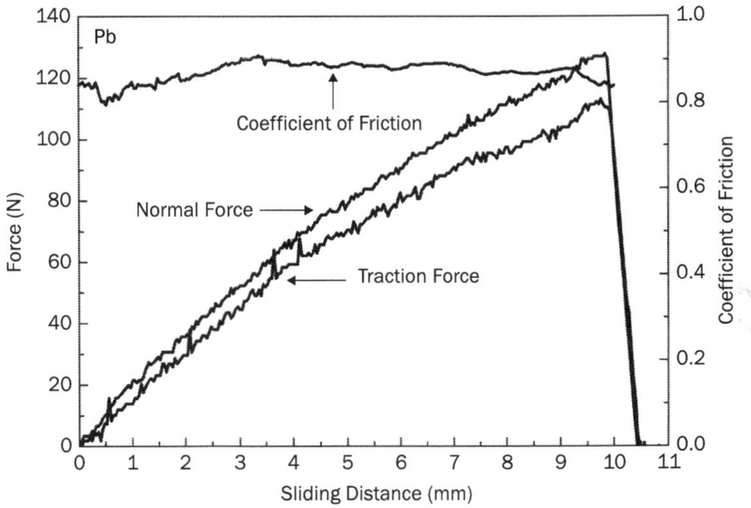

textures. Figure 7.3 shows a typical variation in normal and traction forces with sliding distance obtained in the inclined pin-on-plate experiments. The recording is for the Pb pin slid under dry conditions perpendicular to the unidirectional grinding marks on the steel plate. It is important to note that during the test the normal load was varied with sliding distance and reached to ≈120 N for a sliding distance of 10 mm. The variation of coefficient of friction with sliding distance calculated from these normal and traction forces using Eq. [7.2] is also presented in the same figure. In all the cases, it was observed that the coefficient of friction did not vary significantly for normal loads of up to 120 N (or sliding distance of 10 mm).

Figure 7.4 shows the variation of coefficient of friction with sliding distance when Pb pins slid against steel plates of different surface textures under both dry and lubricated conditions. Here, UPD and UPL represent, respectively, the

Figure 7.4 Variation of coefficient of friction with sliding distance when Pb pins slid against steel plates of different textures under (a) dry and (b) lubricated conditions

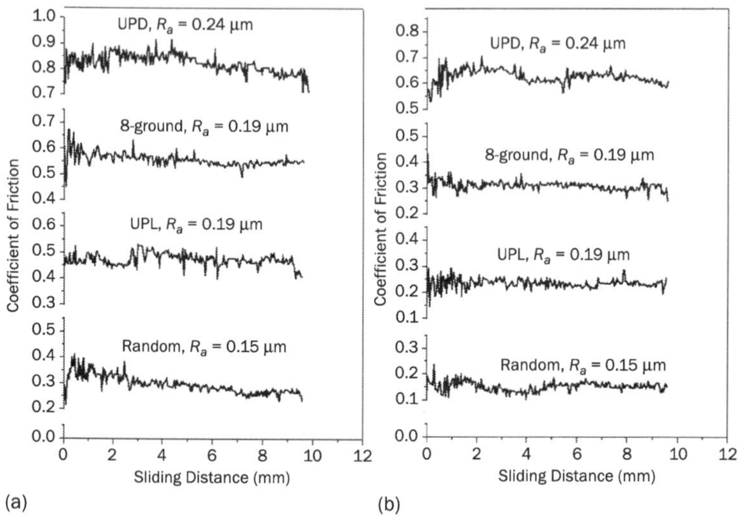

test conditions where the sliding direction is perpendicular and parallel to the unidirectional grinding marks. In the figure, it is observed that the coefficient of friction does not vary significantly with sliding distance. It was observed that the coefficient of friction was relatively high for the UPD surfaces, lower for the 8-ground and the UPL, and finally lowest for the random surfaces under both dry and lubricated conditions. For a given kind of surface texture, experiments were done for different surface roughness (R_a) and it was observed that the coefficient of friction did not vary significantly with surface roughness. The detailed study on the friction and transfer layer formation of Pb material as a function of surface roughness and texture is presented previously (Menezes et al., 2009d). In the previous analysis, the filtered friction data was presented. However, in the present investigation, the unfiltered friction data is considered in order to study the frictional response of different class of materials.

Figure 7.5 shows the variation of coefficient of friction with sliding distance when pure Mg pins were slid against steel plates of different texture under dry and lubricated conditions. Similar to Pb, the coefficient of friction is relatively high for the UPD plates, decreases for the 8-ground and then the UPL plates, and is the lowest for random plates under both dry and lubricated conditions. Stick–slip phenomenon (the oscillation in the coefficient of friction with sliding distance) was observed for the UPD, 8-ground and UPL case under dry conditions, the amplitude of which was highest for the UPD, followed by the 8-ground and least for the UPL case. Zero amplitude of oscillations i.e. no stick–slip phenomenon (steady-state) was observed for the randomly polished steel plates. Under lubricated conditions stick–slip was observed only for the case of UPD and

Figure 7.5 Variation of coefficient of friction with sliding distance when Mg pins slid against steel plates of different textures under (a) dry and (b) lubricated conditions

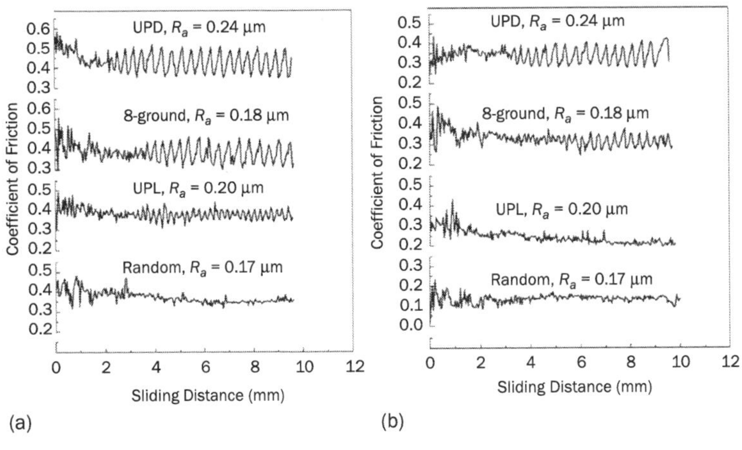

8-ground, where the amplitude of oscillation was greater for the UPD case. No stick–slip phenomenon (steady-state) was observed for the UPL and randomly polished steel plates. For a given kind of surface texture, the roughness of the surface affects neither the average coefficient of friction nor the amplitude of oscillation significantly. The comprehensive analysis on the effect of surface texture and roughness on friction and transfer layer formation of Mg metal is presented elsewhere (Menezes et al., 2006c).

Figure 7.6 shows the variation of coefficient of friction with sliding distance when pure Fe pins slid against steel plates of different surface textures under both dry and lubricated conditions. Akin to Pb, the stick–slip phenomenon was absent for all surface textures under both dry and lubricated conditions. When compared to Pb and Mg, the coefficient of friction for the case of Fe is less dependent on surface textures

Figure 7.6 Variation of coefficient of friction with sliding distance when Fe pins slid against steel plates of different textures under (a) dry and (b) lubricated conditions

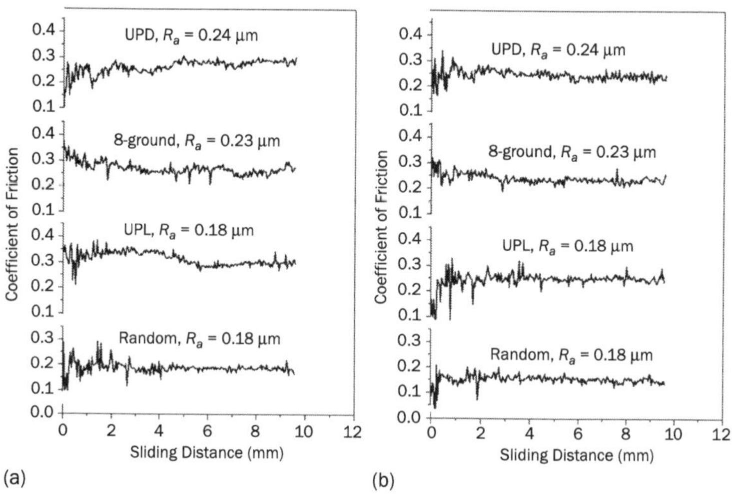

and is relatively high for the UPL surfaces, lower for the 8-ground and then the UPD plates, and lowest for the randomly polished plates under both dry and lubricated conditions. For a given kind of surface texture, the coefficient of friction did not vary significantly with surface roughness. The detailed study on the friction and transfer layer formation of pure Fe as a function of surface roughness and texture has previously been presented (Menezes et al., 2008c). In the previous analysis, the filtered friction data was taken into account, however, in the present case, the unfiltered friction data is considered.

It is understood from the above results that the tribological response of materials during sliding depends on surface texture and to a lesser extent on surface roughness. Hence, the range in which the average coefficient of friction and roughness values fall for the various surfaces under both the dry and lubricated conditions is presented in Figs. 7.7, 7.8 and 7.9 for Pb, Mg and Fe materials, respectively. The error

Figure 7.7 Variation of friction with surface texture for the case of Pb

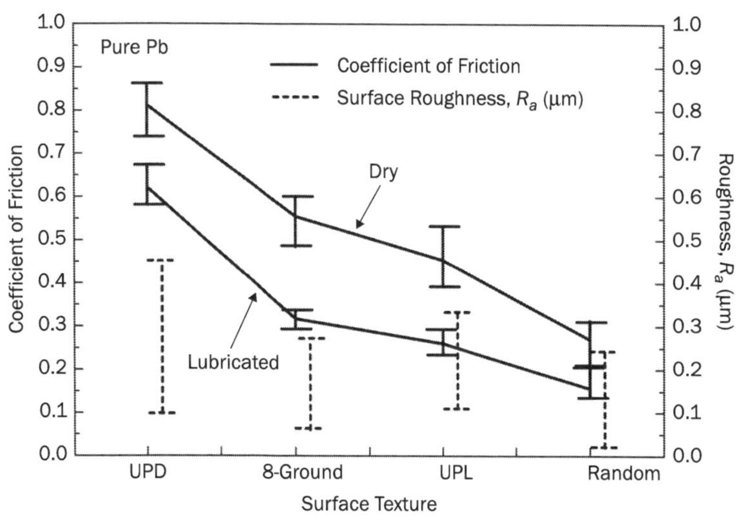

Figure 7.8 Variation of friction with surface texture for the case of Mg

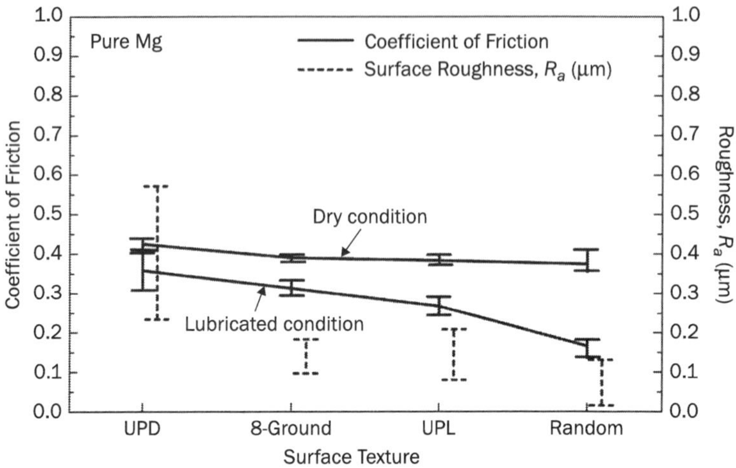

Figure 7.9 Variation of friction with surface texture for the case of Fe

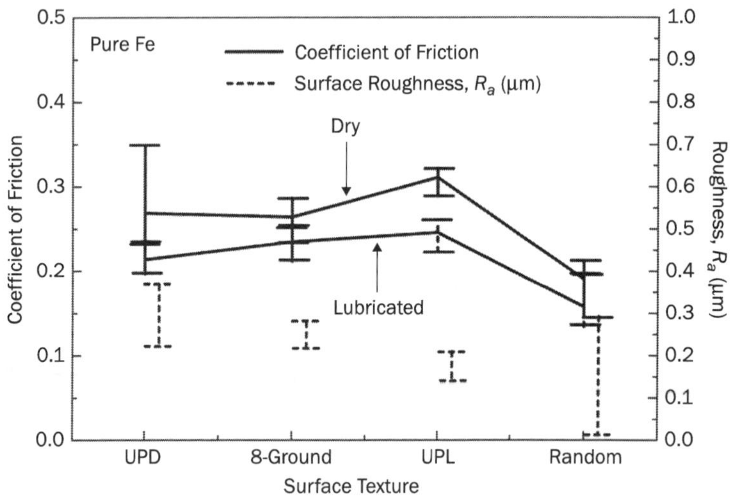

bars in the figures indicate the variation between the maximum and minimum values of the coefficient of friction for the five surface roughnesses. The connecting line refers to the average coefficient of friction of five roughnesses of the same texture. It was observed that the range of surface roughness, R_a, varies between 0.02 and 0.9 μm for different kinds of surfaces. For a given kind of surface texture, it was seen that the coefficient of friction did not vary much with R_a (in the present test range). From the figures, it can be observed that the coefficient of friction varies considerably with surface texture for softer materials like Pb and Mg under both dry and lubricated conditions. However, for the case of harder material like Fe, the coefficient of friction is less dependent on surface texture. More specifically, for softer materials, the coefficient of friction is relatively high for the UPD surfaces, followed in magnitude by the 8-ground and then the UPL, and finally the random surfaces. However, for harder material, the coefficient of friction is relatively high for the UPL surfaces, followed by the 8-ground and then the UPD, and least for the random surfaces.

7.3.2 Variation of transfer layer formation with surface texture

Figures 7.10 (a), (b), (c) and (d) show backscattered scanning electron micrographs of the steel plate surface tested under dry conditions when Pb pins slid on the UPD, 8-ground, UPL and random surfaces, respectively. Here, it can be observed that a large amount of transfer layer of Pb forms on the steel plate surface under dry conditions. It is observed that the amount of transfer layer formed on a steel plate surface is highest for the UPD case followed by the 8-ground plates, the UPL case and then least for the randomly polished steel plates. Figures 7.10 (e), (f), (g) and (h) show the corresponding backscattered scanning electron micrographs of the steel plate surface under lubricated conditions. As expected, it was observed that the amount of transfer layer formed on the steel plates decreases with the application of

Figure 7.10 Backscattered scanning electron micrographs of steel plates when Pb pins slid on UPD (a, e), 8-ground (b, f), UPL (c, g) and random (d, h) textures under dry (a, b, c, d) and lubricated (e, f, g, h) conditions. The arrows indicate the sliding direction of the pin relative to the plate

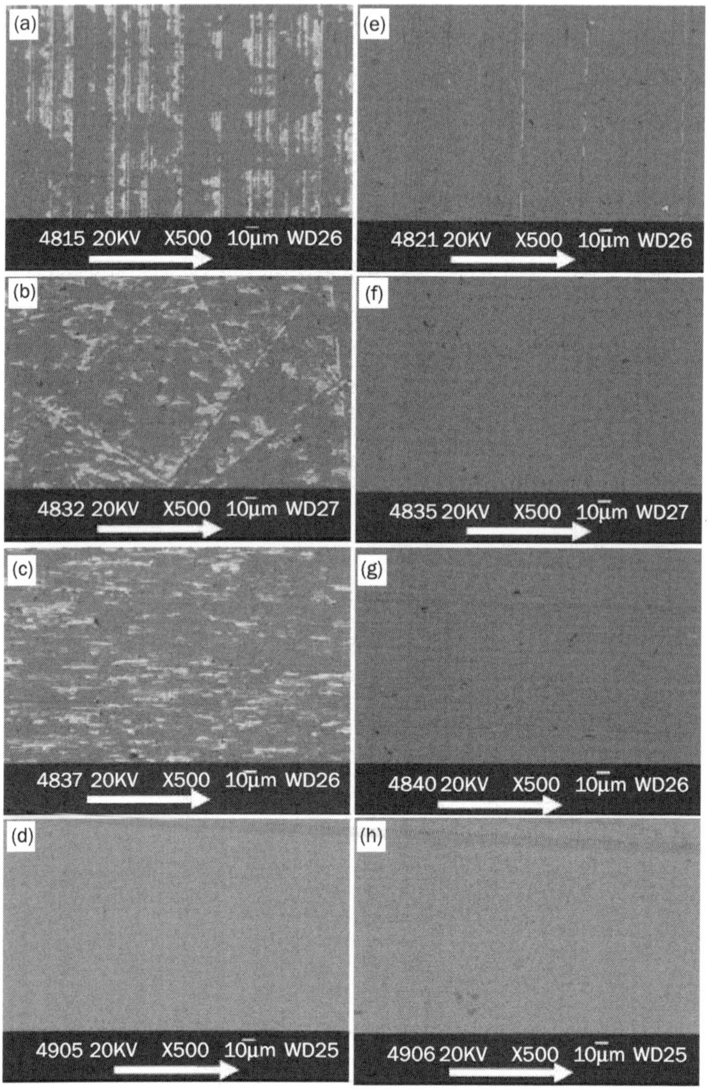

lubricant. In addition, under conditions of lubrication, it was found that the amount of transfer layer formed on the steel plate surface was highest for the UPD case followed by the 8-ground plates, the UPL case and least for the randomly polished steel plates. Under both dry and lubricated conditions, it was observed that the amount of transfer layer formed on the steel plate increases as the normal load increases. For a given kind of surface texture, it was also determined that the amount of the transferred layer formed on the steel plate did not substantially vary with the surface roughness.

A similar observation was made when Mg pins slid on steel plates of different surface textures as shown in the Fig. 7.11. With regard to Fe, Figs. 7.12 (a), (b), (c) and (d) show scanning electron micrographs of the steel plate surfaces tested under dry conditions for UPD, 8-ground surface, UPL surface and random surfaces, respectively. It can be seen that severe damage on the steel plate surface under dry sliding conditions takes place. The damage formed on the plate is much higher for UPD plates and is reduced for 8-ground plates, UPL plates and is least for randomly polished steel plates. For the UPD case, it can be seen that the unidirectional grinding marks, which are perpendicular to the sliding direction, are wiped out during sliding (Fig. 7.12 (a)). However, for the UPL case, a few grinding marks parallel to the sliding direction are wiped out (Fig. 7.12 (c)). For the random case, unidirectional markings along the sliding direction are formed on the steel plate during sliding (Fig. 7.12 (d)). Figures 12 (e), (f), (g) and (h) show the corresponding scanning electron micrographs of the steel plate surfaces under lubricated conditions. It can be observed that the amount of damage formed on the steel plates decreases with the application of lubricant. Similar observations can be made for the plates of similar textures with different roughness values.

Figure 7.11 Backscattered scanning electron micrographs of steel plates when Mg pins slid on UPD (a, e), 8-ground (b, f), UPL (c, g) and random (d, h) textures under dry (a, b, c, d) and lubricated (e, f, g, h) conditions. The arrows indicate the sliding direction of the pin relative to the plate

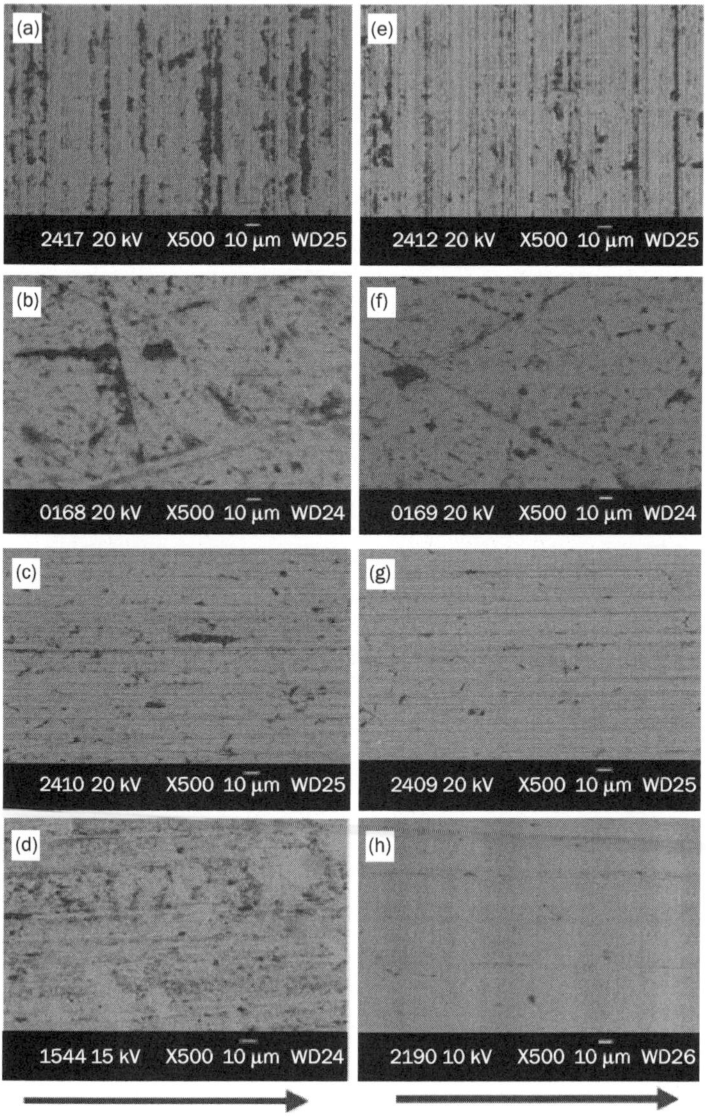

Figure 7.12 Scanning electron micrographs of steel plates when Fe pins slid on UPD (a, e), 8-ground (b, f), UPL (c, g) and random (d, h) textures under dry (a, b, c, d) and lubricated (e, f, g, h) conditions. The arrows indicate the sliding direction of the pin relative to the plate

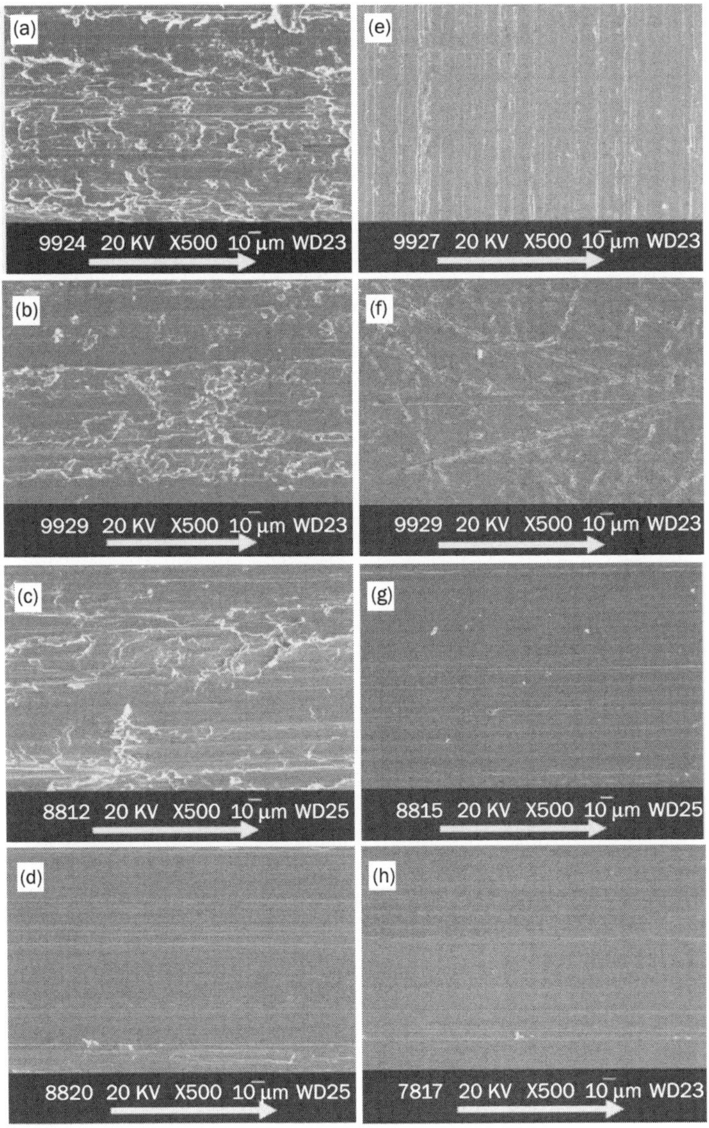

7.4 Discussion

7.4.1 Analysis of friction

When a surface is ground unidirectionally, it will have a 'wave' like texture and a 'hill-and-valley' texture if randomly polished (Menezes et al., 2006b). Considering the present set of experiments, a representative model of a single asperity can be used to describe the physical phenomena involved. More specifically, the interaction can be represented by a softer asperity flowing over harder asperities. When soft materials such as Pb or Mg slide perpendicularly to the unidirectional grinding marks, the softer material have to flow (climb) over the harder asperities of the steel plate. This would induce a higher level of stress, under more pronounced plane strain conditions, leading to severe shear failure and higher material transfer. As higher shear stresses are expected in the UPD test, the coefficient of friction would be large. In the UPL case, the softer materials do not climb over the asperities, and instead flow along the valleys of the steel plate, which requires less energy for the deformation. Thus, the level of stresses and the coefficient of friction generated in UPL tests were lower than those in the UPD tests. For 8-ground surface, softer pins meet the asperities of the steel plate that are aligned in many orientations. One can expect generation of moderate shear stresses, and corresponding friction coefficient for the 8-ground surfaces. For the random surfaces which have a 'hill-and-valley' texture, the softer material can flow around the asperities as the number of asperities opposing the direction of sliding is lower and which leads to smaller stresses and a stress state that is more plane stress near the surface, causing a mild shear failure and lower amount of material transfer.

For the case of harder material such as Fe, the interaction is similar to the interaction between asperities of similar material (both pins and plates are made of iron). When the pin slid perpendicular to the unidirectional grinding marks (UPD), the pin would break the asperities of the steel plate more easily than when the pin slid parallel to the unidirectional grinding marks (UPL). In the UPL case, the asperities are aligned parallel to the sliding direction, and it is difficult to break the asperities. The width of the asperities along the sliding direction in the UPD case is lower when compared to the UPL case. Thus, the UPD requires lower energy during sliding when compared to the UPL. Thus, the friction force generated in the UPD tests was lower than that in the UPL tests. For 8-ground surface, the pin meets the asperities of the steel plate that are aligned in many orientations, and thus generation of a moderate coefficient of friction can be expected. For the random surface, the number of asperities opposing the direction of sliding is lower and thus the coefficient of friction is lower. Thus, the coefficient of friction would be higher for sliding parallel to the grinding marks, and lower for sliding on the random surface. For the 8-ground and unidirectional perpendicular sliding, the coefficient of friction falls between the above two extremes.

Thus, when softer material (Pb and Mg) slides against harder material (steel plate), transfer layer of softer material forms on the harder material under both dry and lubricated conditions as revealed in the backscattered mode using scanning electron microscope. However, when harder material (Fe) slides against harder material (steel plate), the harder pin material damages the surfaces of the harder steel plates and hence the coefficient of friction is less dependent on surface textures. Thus, the variation of coefficient of friction with surface texture is also dependent on hardness of the pin material. A large variation in friction with surface texture was observed for the Pb material which has the lowest hardness (i.e., $8HV_{0.1}$). The variation in friction with surface texture decreases for Mg material which has the moderate hardness (i.e., $55HV_{0.1}$). The variation in friction with surface texture was least for Fe material which has the highest hardness (i.e., $148HV_{0.1}$).

Different frictional responses were observed for the HCP metal (Mg) than the FCC metal (Pb) under similar testing conditions. For the case of Mg, the stick–slip phenomenon was observed under both dry and lubricated conditions. However, no stick–slip phenomenon was observed under both dry and lubricated conditions for the case of Pb. When comparing the properties between Pb and Mg, the major difference is the number of slip systems. Mg has a lower number of slip systems when compared to Pb. A lower number of slip systems can promote stick–slip motion since some of the grains in contact at the asperity level would resist deformation if they were in the preferred orientation (Buckley and Johnson, 1968; Farhat, 2001). Comparing the magnitude of stick–slip oscillations, the lubricated case was lower than the unlubricated case for the Mg pins. When friction decreases under lubricated conditions, the extent of asperity interlocking also decreases which reduces the amplitude of stick–slip motion. It was also observed that the amplitude of oscillation for Mg did not vary with normal load once it reached a critical value. This can be attributed to the fact that once the maximum real area of contact is achieved, further increase in normal load does not cause any change in the real area of contact. Previous efforts also showed that the stick–slip phenomenon was absent for Cu pins (Menezes et al., 2009c) and the stick–slip phenomenon was present for the Zn (Menezes et al., 2008b) pin during sliding against steel plate of different surface textures under identical testing conditions. This clearly shows that FCC materials exhibit steady-state sliding and HCP materials stick–slip sliding.

It was stated earlier that the 8-ground surfaces have the asperities which are aligned in many orientations. The coefficient of friction values obtained for the 8-ground surfaces fell in between those of UPD and UPL surfaces. Hence, efforts were made earlier by the authors (Menezes et al., 2010) to study the influence of grinding marks orientation on friction and transfer layer formation during sliding. In the experiments, unidirectional grinding marks were created on the steel plates. Then, pins made of pure Al, Al-4Mg alloy and pure Mg were slid at angles of 0° (parallel), 5°, 10°, 15°, 20°, 25°, 45°, 65°, 80°, 85° and 90° (perpendicular) to the unidirectional grinding marks on the steel plate under both dry and lubricated conditions. It was observed that the average coefficient of friction and transfer layer formation depend on the directionality of the grinding marks but were less dependent of surface roughness on the harder mating surface. In the present set of experiments, for the 8-ground surfaces, the grinding marks orientation lie in between the UPD and UPL surfaces and hence the 8-ground surfaces will then be expected to have a friction between that of UPD and UPL surfaces as is the case observed in the present work and previous work (Menezes et al., 2010).

In this study, the tribological behavior is explained in terms of the constrained nature induced by the harder asperities to the softer asperities during sliding. When the constraint offered by the harder asperities is greater, a higher amount of energy is needed to overcome the frictional resistance offered by the surface. This induces higher levels of stresses in the deforming material and higher amounts of transfer layer formation on the harder counter-surface. Previous work carried out on dies with various surface textures did not differentiate between the surfaces that offer various levels of constraint to flow. The work done by Lakshmipathy and Sagar (1992) and Costa and Hutchings (2009) showed that the friction factor is lower for those surfaces that act as lubricant reservoirs and induce hydrodynamic effects. They, however, did not differentiate the role of constraint to flow that results in the reduction of friction factor. Recently, Ma and Lu (2011) prepared similar kinds of surface textures to those used in the present investigation and the effect of surface texture of harder material on friction and transfer layer formation during sliding against copper–graphite composite was studied. They found that coefficient of friction and transfer layer formation are dependent on the surface texture of the harder counter–surfaces. Ma and Lu (2011) also reported that the formation of continuous and discontinuous transfer layer on the harder counter–surface is dependent on the variations in surface textures and this is attributed to the gradient of deformation owing to the height of the harder asperities.

In was reported in the literature (Wagner, 2001) that the tool surface texture and sheet metal surface texture influence frictional behavior during metal forming. In this study, only the surface texture of a harder counter–surface is considered. The initial surface texture of both harder and softer materials is important in controlling the frictional behavior. However, the authors believe that the harder surface texture plays a crucial role in controlling the frictional phenomenon when compared to the softer surface texture. The harder surface asperities do not experience deformation while the softer asperities do experience deformation during sliding.

7.4.2 Applications

The results achieved in this study can be applied to control the coefficient of friction at the interface between the die and sheet metal in metal forming process. It is possible to achieve a particular coefficient of friction at a particular area at the interface between the die and sheet metal by designing a proper surface texture on the die material. To attain a high value of coefficient of friction, the surface texture on the die can be designed such that the grinding marks are perpendicular to the flow direction. On the other hand, to obtain a low coefficient of friction, random surface texture may be designed. It can also reduce the amount of flash formation and folding during forging when appropriate friction is given. This would lead to significant saving of material. Controlling textures may also enable the reduction of conventional lubricants. In practical situations, alloys of the present set of materials are commonly used in metal forming processes. However, in this study, efforts were made to understand the fundamental aspects of friction using pure materials that decrease the number of variables.

The work presented in this study was confined to a single sliding event; however, in metal forming processes, the dies can be used for multiple sliding events. The surface texture of the die and thus the friction might be changed substantially due to previous sliding events where the original surface texture of the die might self-organize and produce a new surface texture according to the sliding conditions. By applying suitable coatings to preserve the original surface, textures could retain the initial friction values and thus the life of the die.

7.5 Conclusions

In the present investigation, experiments were conducted using an inclined pin-on-plate apparatus to study the tribological response of materials with varying hardness and crystal structure during sliding on various surface textures. The investigation was carried out using pins made of Pb, Mg and Fe against EN8 steel plates. The conclusions based on the experimental results are as follows:

- The coefficient of friction is controlled by the surface texture of the harder mating surfaces.
- The variation in friction with surface texture is dependent on hardness of the pin materials.
- The variation is highest for Pb and decreases for Mg and is least for Fe materials.
- Two kinds of friction response, namely steady-state and stick–slip, were observed during sliding. FCC (Pb) and Fe (BCC) materials exhibited steady-state response and HCP (Mg) exhibited stick–slip response.
- In general, the frictional response was dependent on material pair, normal load, lubrication and surface texture of the harder material.
- The transfer layer was observed when Pb and Mg pins slid against steel plates of different surface textures. However, no transfer layer was observed for the case of Fe and the Fe pins abrade all surfaces during sliding.

7.6 References

Bello D O and Walton S (1987) 'Surface topography and lubrication in sheet–metal forming', *Tribol Int*, **20**(2): 59–65.

Bhushan B (2002) *Introduction to Tribology*. New York: John Wiley and Sons Inc.

Bouissou S, Petit J P and Barquins M (1998) 'Normal load, slip rate and roughness influence on the polymethylmethacrylate dynamics of sliding: 1. Stable sliding to stick–slip transition', *Wear*, **214**(2): 156–64.

Bowden F P and Tabor D (1950) *The Friction and Lubrication of Solids, Part 1*. Oxford: Clarendon Press.

Buckley D H (1967) 'Friction characteristics in vacuum of single and polycrystalline aluminum oxide in contact with themselves and with various metals', *ASLE Trans*, **10**: 134–45.

Buckley D H and Johnson R L (1968) 'The influence of crystal structure and some properties on hexagonal metals on friction and adhesion', *Wear*, **11**(6): 405–19.

Costa H L and Hutchings I M (2009) 'Effects of die surface patterning on lubrication in strip drawing.' *J Mater Process Technol*, **209**(3): 1175–80.

Don J, Sun T C and Rigney D A (1983) 'Friction and wear of Cu–Be and dispersion-hardened copper systems', *Wear*, **91**(2): 191–9.

Dwivedi D K, Arjun T S, Thakur P, Vaidya H and Singh K (2004) 'Sliding wear and friction behaviour of Al–18% Si–0.5% Mg alloy', *J Mater Process Technol*, **152**: 323–8.

Farhat Z N (2001) 'Contribution of crystallographic texturing to the sliding friction behaviour of FCC and HCP metals', *Wear*, **250**(1–12): 401–8.

Hayashi N, Matsui A and Takahashi S (1999) 'Effect of surface topography on transferred film formation in plastic and metal sliding system', *Wear*, **225–9**: 329–38.

Hu Z M and Dean T A (2000) 'A study of surface topography, friction and lubricants in metal forming', *Int J Mach Tool Manufact*, **40**(11): 1637–49.

Hwang D H and Gahr K H Z (2003) 'Transition from static to kinetic friction of unlubricated or oil lubricated steel/steel, steel/ceramic and ceramic/ceramic pairs', *Wear*, **255**(1–6): 365–75.

Kayaba T (1962) 'A study of the wear and friction of some bearing materials', *Wear*, **5**: 173–81.

Koura M M (1980) 'The effect of surface texture on friction mechanisms', *Wear*, **63**(1): 1–12.

Kudo H (1965) 'A note on the role of microscopically trapped lubricant at the tool–work interface', *Int J Mech Sci*, 7(5): 383–8.

Lakshmipathy R and Sagar R (1992) 'Effect of die surface topography on die-work interfacial friction in open die forging', *Int J Mach Tool Manufact*, **32**(5): 685–93.

Lim S C, Ashby M F and Brunton J H (1989) 'The effects of sliding conditions on the dry friction of metals', *Acta Metall*, **37**: 767–72.

Lovell M R, Kabir M A, Menezes P L and Higgs III C F (2010) 'Influence of boric-acid additive size on green lubricant performance', *Phil Trans R Soc A*, **368**: 4851–68.

Ma W and Lu J, (2011) 'Effect of surface texture on transfer layer formation and tribological behaviour of copper–graphite composite', *Wear*, **270**: 218–29.

Määttä A, Vuoristo P and Mäntylä T (2001) 'Friction and adhesion of stainless steel strip against tool steels in unlubricated sliding with high contact load', *Tribol Int*, **34**(11): 779–86.

Malayappan S and Narayanasamy R (2004) 'An experimental analysis of upset forging of aluminium cylindrical billets considering the dissimilar frictional conditions at flat die surfaces', *Int J Adv Manuf Technol*, **23**: 636–43.

Menezes P L, Kishore and Kailas S V (2006a) 'Effect of roughness parameter and grinding angle on coefficient of friction when sliding of Al-Mg alloy over EN8 steel', *ASME J Tribol*, **128**(4): 697–704.

Menezes P L, Kishore and Kailas S V (2006b) 'Studies on friction and transfer layer using inclined scratch', *Tribol Int*, **39**(2): 175–83.

Menezes P L, Kishore and Kailas S V (2006c) 'Influence of surface texture on coefficient of friction and transfer layer formation during sliding of pure magnesium pin on 080 M40 (EN8) steel plate', *Wear*, **261**(5–6): 578–91.

Menezes P L, Kishore, Shimjith M and Kailas S V (2007) 'Influence of surface texture on friction and transfer layer formation in Mg-8Al alloy/steel tribo-system', *Indian J Tribol*, 2(1), 46–54.

Menezes P L, Kishore and Kailas S V (2008a) 'Effect of surface topography on friction and transfer layer during sliding', *Tribol Online*, 3(1): 25–30.

Menezes P L, Kishore and Kailas S V (2008b) 'Role of surface texture and roughness parameters in friction and transfer layer formation under dry and lubricated sliding conditions', *Int J Mater Res*, **99**(7): 795–807.

Menezes P L, Kishore and Kailas S V (2008c) 'Studies on friction in iron-steel tribo-system under dry and lubricated conditions', *Mater Manuf Process*, **23**(7): 698–707.

Menezes P L, Kishore and Kailas S V (2009a) 'Influence of inclination angle of plate on friction, stick–slip and transfer layer – a study of magnesium pin sliding against steel plate', *Wear*, **267**(1–4): 476–84.

Menezes P L, Kishore and Kailas S V (2009b) 'Influence of surface texture and roughness parameters on friction and transfer layer formation during sliding of aluminium pin on steel plate', *Wear*, **267**(9–10): 1534–49.

Menezes P L, Kishore and Kailas S V (2009c) 'Study of friction and transfer layer formation in copper–steel tribo-system', *Tribol Trans*, **52**(5): 611–22.

Menezes P L, Kishore, Kailas S V and Lovell M R (2009d) 'Influence of roughness parameters and surface texture on friction during sliding of pure lead over 080 M40 steel', *Int J Adv Manuf Technol*, **43**(7–8): 731–43.

Menezes P L, Kishore, Kailas S V and Lovell M R (2010) 'Response of materials as a function of grinding angle on friction and transfer layer formation', *Int J Adv Manuf Technol*, **49**(5): 485–95.

Menezes P L, Kishore, Kailas S V and Lovell M R (2011) 'Influence of inclination angle and machining direction on friction and transfer layer formation', *J Tribol*, **133**(1): 014501.1–014501.8.

Mokhtar M O A, Zaki M and Shawki G S A (1979) 'Effect of mechanical properties on frictional behaviour of metals', *Tribol Int*, **12**: 265–8.

Mokhtar M O A, Zaki M and Shawki G S A (1980) 'Correlation between the frictional behaviour and the physical properties of metals', *Wear*, **65**: 29–34.

Nellemann T, Bay N and Wanheim T (1977) 'Real area of contact and friction stress – the role of trapped lubricant', *Wear*, **43**(1): 45–53.

Rasp W and Wichern C M (2002) 'Effects of surface-topography directionality and lubrication condition on frictional behaviour during plastic deformation', *J Mater Process Technol*, **125–26**: 379–86.

Saha P K, Wilson W R D and Timsit R S (1996) 'Influence of surface topography on the frictional characteristics of 3104 aluminum alloy sheet', *Wear*, **197**(1–2): 123–9.

Schedin E (1994) 'Galling mechanisms in sheet forming operations', *Wear*, **179**(1–2): 123–8.

Staph H E, Ku P M and Carper H J (1973) 'Effect of surface roughness and surface texture on scuffing', *Mech Mach Theor*, **8**: 197–208.

Suh N P (1986) *Tribophysics*, New Jersey: Prentice-Hall, Inc.

Wagner S (2001) 'Optimizing friction between die and sheet metal', *Proceedings of Global Symposium on Innovations in Materials, Processing and Manufacturing*, United States, 245–67.

Wakuda M, Yamauchi Y, Kanzaki S and Yasuda Y, (2003) 'Effect of surface texturing on friction reduction between ceramic and steel materials under lubricated sliding contact', *Wear*, **254**(3–4): 356–63.

8

Laser straight cutting of Ti-6Al-4V alloy: Temperature and stress fields

B. Yilbas and S. Akhtar, King Fahd University of Petroleum and Minerals, Saudi Arabia, and O. Keles, Gazi University, Turkey

Abstract: Laser straight cutting of titanium alloy is carried out. Thermal stress fields in the cutting section are examined using the finite element numerical code. The morphological changes in the laser cutting region are investigated using optical microscopy, scanning electron microscopy and *x*-ray diffraction technique. It is found that von Mises stress attains high values in the vicinity of the top and bottom surfaces as well as at the mid-thickness of the workpiece. Some small sideways burns in the cut section are observed, which are associated with the high temperature oxidation variations during the cutting process. Due to the brittle structure in the re-cast layer, a few locally scattered microcracks are observed at the re-cast layer surface.

Key words: laser, cutting, titanium alloy, thermal stress.

8.1 Introduction

Laser cutting of titanium alloys requires proper setting of cutting parameters as titanium has a high affinity to oxidation reactions. High temperature oxidation reactions result in sideways burning around the cut edges while lowering the cutting quality. To eliminate/minimize high oxidation reactions, an inert assisting gas is used to shield the cutting section from oxygen, which in turn suppresses high temperature oxidation reactions during the cutting process. The assisting gas has twofold effects, which are i) shielding the section from oxygen, and ii) convection cooling of the cutting section. Since the assisting gas is at high pressure, convection cooling results in high cooling rates in the cutting section while forming a brittle and dense layer at the kerf surface; in which case, thermal stress levels become high in this region. The model study provides information on the physical processes taking place during the cutting process. Consequently, investigation of the thermal stress field in the vicinity of the cut section of titanium alloy becomes essential.

Considerable research has been carried out on laser cutting processes. Laser melting of titanium and other alloys in relation to machining applications was examined by Shen et al. (2001). They presented the temporal variation of the melt profile for various metals. Laser heating of Ti-6Al-4V alloy was investigated by Yang et al. (2010). They indicated that the depth and width of the heat affected zone increased with an increase of the laser spot size as well as of the laser scanning speed. Laser cutting of titanium alloy sheet was examined by Deng et al. (2005). The findings revealed that argon as an assisting gas provided a shielding effect in the cutting section while reducing the oxidation reactions. Laser cutting of titanium with pulsed Nd: YAG laser was carried out by Maher and Tong (1998). They showed that this inert assisting gas resulted in good quality cut edges with minimum sideways burning. Laser cutting of memory shape alloys was studied by Pfeifer et al. (2010). Their findings showed that due to the local energy input, a smaller-sized heat affected zone was formed and the alloy memory properties remained the same in the laser cutting section.

Statistical analysis for laser cutting of titanium was carried out by Almeida et al. (2006). They introduced a factorial analysis for different processing parameters and employed the statistical tests to identify the significant levels of each factor. Laser cutting of titanium alloy sheet was carried out by Shanjin and Yang (2006). They examined the cut section morphology for different assisting gases. Inert gas cutting of titanium sheet with pulsed mode CO_2 laser was examined by Rao et al. (2005). They showed that laser cutting with high frequency and low duty cycle pulse modes produced dross-free cuts with an unnoticeable heat affected zone. The influence of process parameters during laser cutting of nickel-based superalloys was investigated by Thawari et al. (2005). They presented cut section quality in terms of the cut edges. Yilbas et al. (2011) studied laser cutting of holes in titanium alloys. The findings revealed that stress levels around the cut section increased for small hole diameters.

In the present study, laser straight cutting of Ti-6Al-4V alloy is carried out. Temperature and stress fields developed in the cutting section are predicted using ABAQUS finite element code (2009). The morphological changes in the cut section and microstructural changes in the kerf vicinity are examined using the optical and scanning electron microscopes. The residual stress developed in the vicinity of the cut surface is obtained using the X-ray diffraction (XRD) technique and the residual stress data is compared with its counterpart predicted from numerical simulations.

8.2 Heating and stress analysis

Figure 8.1 shows the schematic view of the laser cutting and the coordinate system. The simulation for heating and stress analysis is performed in ABAQUS/Standard and consists of sequential thermal stress analysis. In the sequential thermal stress analysis, 142,468 elements and 132,140 hexahedral elements are used. Heat transfer analysis used mesh elements of type DC3D8 (8-node linear heat transfer brick) and stress analysis used C3D8 elements (8-node linear stress brick). Since the melting is modeled through a temperature-dependent specific heat capacity allowing the latent heat of melting and solidus and liquidus temperatures of the material, the solid–liquid interface can be determined from the enthalpy balance (Yilbas et al., 2011). The temperature data are transferred to the elements used for the stress analysis through the connectivity matrix. This required less computational time for the converged results. The fixed boundary conditions are applied on the both ends of the workpiece resembling the experimental laser heating situation.

Figure 8.1 A schematic view of laser cutting of workpiece and coordinate system

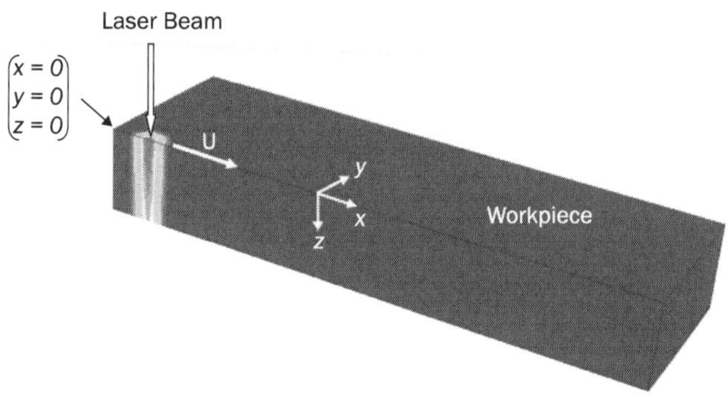

In the stress analysis, displacements are stored by ABAQUS at the nodal positions as a solution variable, and loads are defined as prescribed displacements and forces. Employing the interpolation functions, it is possible to calculate the strain and stress increments at any point within the element using the compatibility and constitutive equations. ABAQUS transforms the mechanical equilibrium equations into a set of simultaneous equations, such that the nodal displacements and forces are related to each other through the elemental stiffness matrix.

Laser heat flux with Gauss distribution and prescribed velocity of 10 cm/s along the x-axis through user subroutine DFLUX is applied to the thermal model. The Gauss parameter a is a = 0.0003 m, in accordance with the experimental power intensity distribution. The thermal model consisted of two steps. The first step, which lasts 0.05 seconds, simulates the response of the plate under moving laser heat flux. The second step, which lasts for 1000 seconds, simulates the continued cooling in the model. Cooling was allowed to continue until all of the plate reaches initial temperature (room temperature). The temperature–time history resulting from the thermal analysis is used as input to the thermal stress analysis. The workpiece is considered as an elastic body, which is modeled as von Mises elastic–plastic material with isotropic hardening and with a yield stress that changes with temperature. Table 8.1 gives the properties of Ti-6Al-4V alloy used in the simulations.

8.3 Experimental data

A CO_2 laser (LC-ALPHA III) delivering nominal output power of 2 kW in pulse mode is used to cut the workpiece. The nominal focal length of the focusing lens is 127 mm. The

Table 8.1 Thermo-physical and mechanical properties of Ti-6Al-4V alloy (VERIDIAM, 2010) used in the simulations

Density kg/ m^3	Latent Heat of Melting (J/kg)	Solidus Temperature (K)	Liquidus Temperature (K)	
4430	4.2×10^5	1878	1933	
a) Density and melting properties of Ti-6Al-4V alloy				
Temperature (K)	Cp (J/kgK)	k (W/mK)		
300	403	6.5		
573	505	7		
673	520	9		
773	530	11		
873	545	12		
1073	620	17		
b) Thermal properties of Ti-6Al-4V alloy				
Temperature (K)	Elastic Modulus (GPa)	Poisson's ratio	Expansion Coefficient × 10^6 (1/K)	Yield Stress (MPa)
300	113	0.34	8	14.37
373	108	0.34	8.6	12.81
473	103	0.34	9	11.2
589	95	0.34	9.2	10.5
698	85	0.34	9.4	7
803	70	0.34	9.7	5
973	50	0.34	1.05	2.25
c) Mechanical properties				

laser beam diameter focused at the workpiece surface is 0.25 mm. Nitrogen assisting gas emerging from the conical nozzle and co-axially with the laser beam is used. The laser cutting parameters are given in Table 8.2. Ti-6Al-4V alloy sheets with 3 mm thickness were used as workpieces. A JEOL JDX-3530 scanning electron microscope (SEM) is

Table 8.2 **Laser cutting conditions used in the experiment**

Cutting Speed (cm/s)	Power (W)	Frequency (Hz)	Nozzle Gap	Nozzle Diameter (mm)	Focus Setting (mm)	N2 Pressure (kPa)
10	2000	1000	1.0	1.5	127	600

used to obtain photomicrographs of the cross-section of the workpieces after the tests. The Bruker D8 Advance diffractometer utilizing Cu-Kα radiation is used for XRD analysis. A typical setting of XRD was 40 kV and 30 mA. It should be noted that the residual stress measured using the XRD technique provides the data in the surface region of the specimens.

The residual measurement relies on the stresses in fine grained polycrystalline structure. The relationship between the peak shift and the residual stress (σ) is given (H&M Analytical Services, Inc., 2010 and Khana et al., 2005):

$$\sigma = \frac{E}{(1+\upsilon)sin^2\psi}\frac{(d_n - d_o)}{d_o} \qquad [8.1]$$

where E is Young's modulus, ν is Poisson's ratio, ψ is the tilt angle, d_o is the stress free spacing, and d_n is the d spacing measured at each tilt angle. If there are no shear strains present in the specimen, the d spacing changes linearly with $sin^2\psi$. The calculations for the residual stress were performed for the α-Ti (101) plane of Ti-6Al-4V alloy while using the elastic modulus and Poisson's ratio of Ti-6Al-4V alloy as 113 GPa and 0.34, respectively (VERIDIAM, 2010). The residual stress, therefore, determined from the XRD technique at the surface vicinity is of the order of 720±30 MPa. XRD measurements are repeated three times and the error related to the measurements is of the order of 3%. In addition, the coefficient of determination of the linear fit is $R = 0.97$.

8.4 Results and discussion

Laser nitrogen gas assisted cutting of Ti-6Al-4V alloy sheet is carried out. The morphological and microstructural changes in the cut section are examined using optical and scanning electron microscopes. Temperature and stress fields formed around the cut section are predicted using the finite element code. The residual stress predicted in the vicinity of the cut section is measured using the XRD technique. The residual stress predictions are compared with the XRD data.

Figure 8.2 shows temperature distribution along the x-axis for different cooling periods, while Fig. 8.3 shows temperature contours immediately after the end of the cutting process. The x-axis represents the distance along the laser cut edge. It should be noted that the cooling period starts after the cutting ends, which corresponds to t = 0.08 s in the figure. Since the laser cutting speed is 0.1 m/s, the laser beam is at x = 0.008 m where temperature is well above the melting

Figure 8.2 Temperature distribution along the *x*-axis for different cooling periods

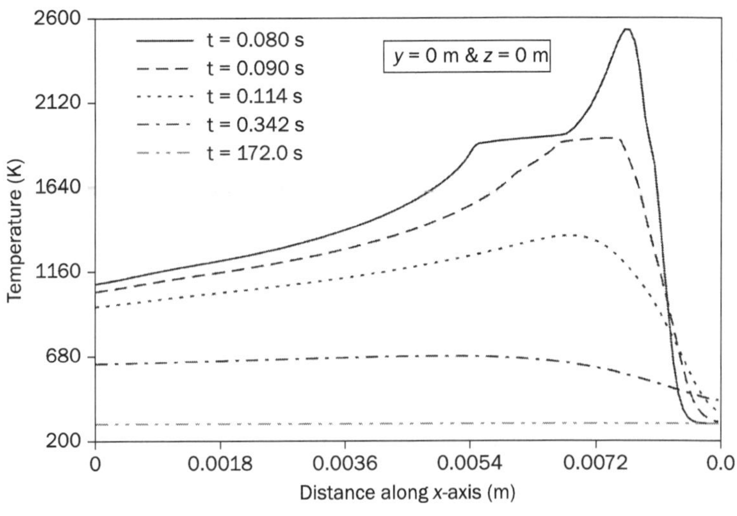

Figure 8.3 Temperature contours in the cutting section at the beginning of the cooling cycle

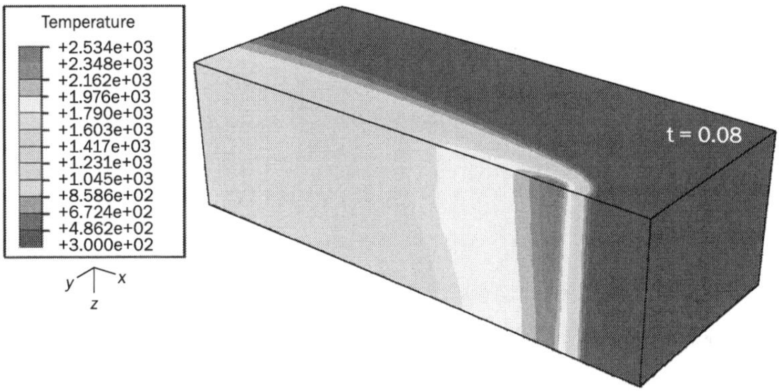

temperature of the substrate material and the laser beam is switched-off at this location. It is evident that behind the laser spot ($x \leq 0.007$ m), temperature remains the same at the melting temperature of the workpiece. This indicates that the cut edge is at melting temperature in this region. However, temperature drops below the melting temperature along the cut edge $0 \leq x \leq 0.005$ m, which is associated with the conduction, convection and radiation cooling of the initially cut edges. Since the thermal diffusivity of the alloy is low, temperature decay is gradual. Moreover, attainment of considerably higher temperature at the irradiated spot indicates the presence of superheating in the liquid phase.

As the cooling period progresses ($t > 0.08$ s), temperature reduces along the cut edges; however, temperature decay in the irradiated spot is gradual due to superheating of the liquid phase. The temperature gradient is high in the front region of the laser irradiated spot ($x > 0.0082$ m). This is attributed to large temperature difference between the irradiated spot and the base material, which is at a low temperature. Consequently, the difference between superheated liquid temperature and low base material temperature results in a high temperature gradient in this region.

Figure 8.4 shows von Mises stress along the *x*-axis for different cooling periods, which are similar to those shown in Fig. 8.2, while Fig. 8.5 shows von Mises stress contours immediately after the end of the cutting process. Von Mises stress attains low values in the regions where temperature is high. This is associated with the temperature-dependent elastic modulus, which reduces with increasing temperature (Table 8.1). Moreover, von Mises stress increases as temperature along the cut edge reduces. In addition, von Mises stress attains high values in the front region of the laser spot where temperature gradient is high. This is more pronounced in the early cooling period (t = 0.08 s). Von Mises stress reduces in the region where superheating

Figure 8.4 Von Mises stress distribution along the x-axis for different cooling periods

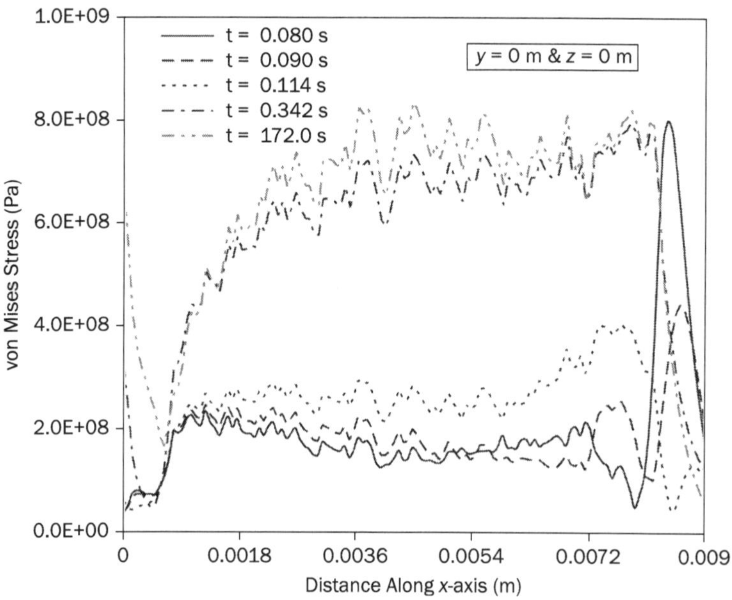

Figure 8.5 Von Mises stress contours in the cutting section at the beginning of the cooling cycle

of liquid takes place at onset of cooling cycle. This is attributed to the self-annealing effect of the laser irradiated region due to high temperature. In this case, slow cooling results in stress relaxation in this region. As the cooling period ends (t = 172 s) and temperature reduces to initial temperature, von Mises stress remains high. The maximum value of von Mises stress along the cut edge is of the order of 800 GPa.

Figure 8.6 shows temperature distribution along the y-axis for different cooling periods. It should be noted that the y-axis is normal to the laser cutting direction at the surface (Fig. 8.1). Temperature remains high in the vicinity of the cut edge and decays sharply as the distance from the cut edge vicinity increases along the y-axis. Moreover, temperature decay in the vicinity of the cut edge is gradual, which can be associated with the superheating of the liquid phase. The gradual decay of temperature results in a low temperature gradient while sharp decay generates a high temperature gradient along the y-axis. Temperature decay rate is lower in the vicinity of the cut edge. This is attributed to occurrence of superheating in liquid phase in this region.

Figure 8.6 **Temperature distribution along the *y*-axis for different cooling periods**

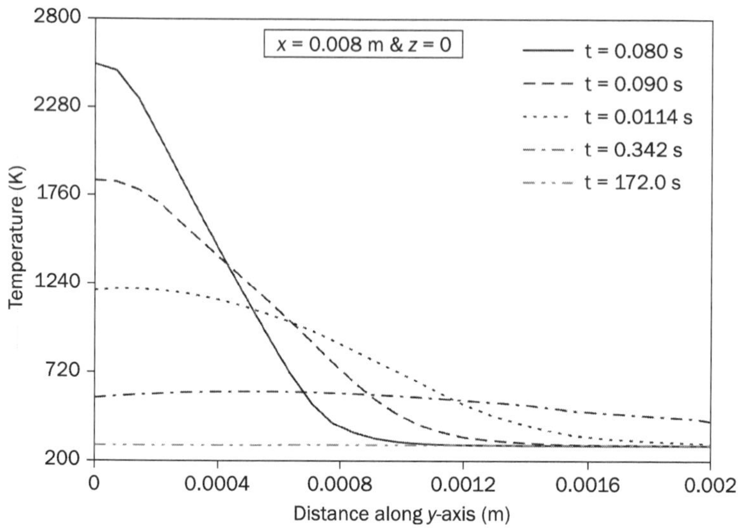

Figure 8.7 shows von Mises stress distribution along the *y*-axis for different cooling periods similar to those shown in Fig. 8.6. Von Mises stress attains low values in the region where temperature is high, particularly in the early period, while it increases in the region of high temperature gradient. This is associated with the elastic modulus of the workpiece (Table 8.1), which reduces with increasing temperature. As the cooling period progresses, the maximum von Mises stress moves towards the edge of the cut section. The maximum von Mises stress is of the order of 750 MPa and high stress region extends 80 m along the *y*-axis, whereas the magnitude of the von Mises stress is of the order of 180 MPa at 80 m and beyond. However, as the distance along the *y*-axis increases further, von Mises stress remains at the same low value.

Figure 8.8 shows temperature distribution along the *z*-axis for different cooling periods. It should be noted that the *z*-axis extends along the thickness of the workpiece.

Figure 8.7 Von Mises stress distribution along the *y*-axis for different cooling periods

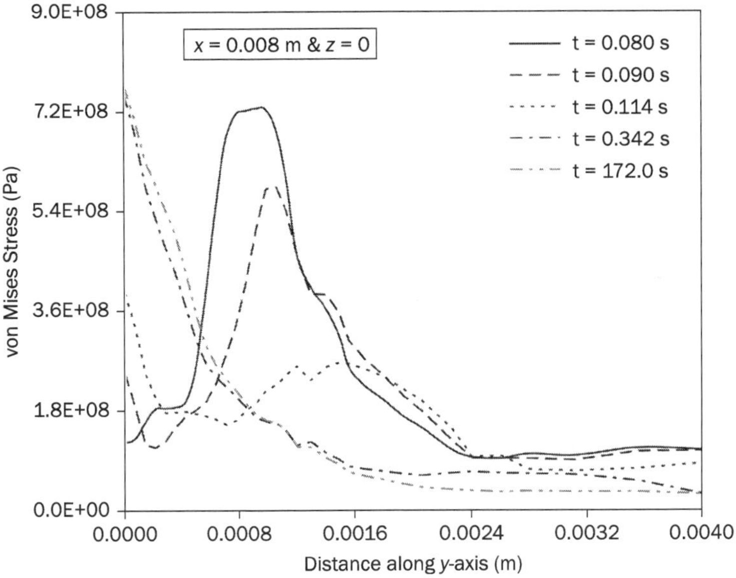

Figure 8.8 Temperature distribution along the *z*-axis for different cooling periods

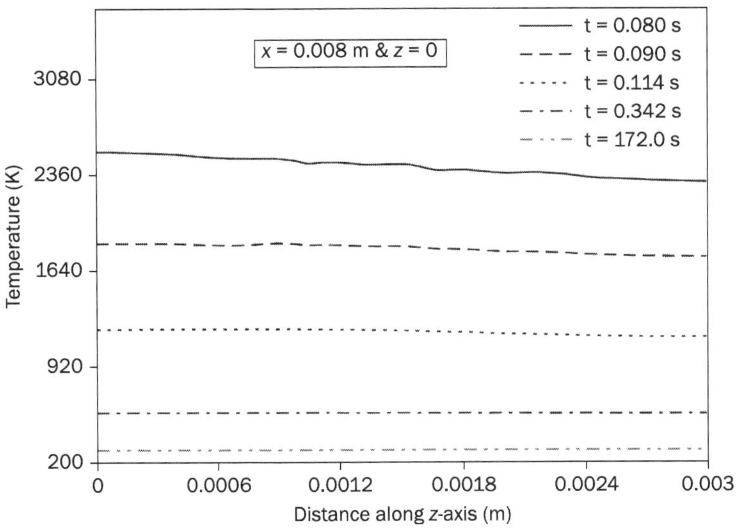

Consequently, z = 0 m represents the free surface where the laser beam interacts with the substrate material during cutting while z = 3 mm corresponds to the back surface of the workpiece. Temperature decays gradually along the thickness of the cut section. The gradual decay of temperature is associated with the cutting surface where temperature remains above the melting temperature of the substrate material. It should be noted that x = 0 and y = 0 correspond to the kerf (cut edge) surface. As time progresses, temperature decays almost uniformly along the thickness at the kerf surface. This is attributed to the convection, conduction and radiation heat losses from the kerf surface.

Figure 8.9 shows von Mises stress distribution along the z-axis for different cooling periods similar to those shown in Fig. 8.8. Von Mises stress attains low values at high temperatures during the initial cooling period and it increases considerably as the cooling period progresses. Von Mises

Figure 8.9 Von Mises stress distribution along the z-axis for different cooling periods

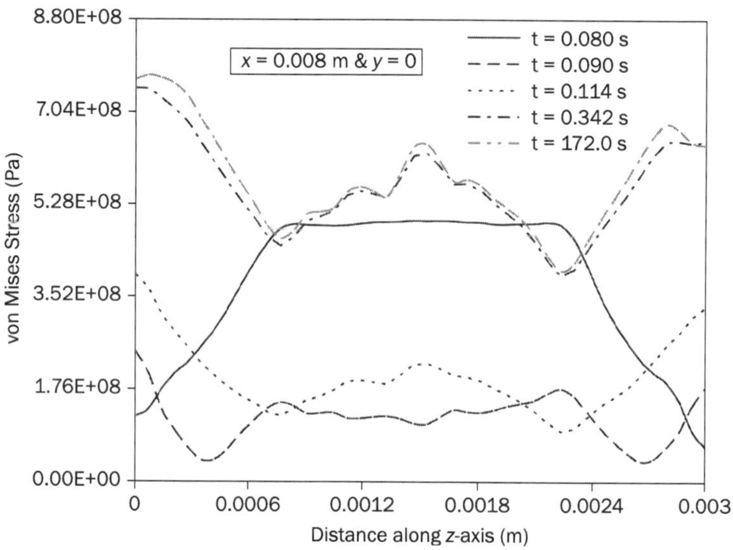

stress reaches its maximum in the vicinity of the top and bottom surfaces of the workpiece, despite the fact that the top and bottom surfaces of the workpiece are free to expand. The attainment of high von Mises stress in these regions is attributed to thermal strain formation. Moreover, von Mises stress increases in the mid-thickness of the workpiece; however, its value is less than the maximum stress. This behavior is associated with the compressive stress which is formed in this region. Since the top and bottom surfaces expand freely, the mid-thickness region undergoes compression due to free expansion in the workpiece surface region.

Figure 8.10 shows optical photographs of the top and bottom surfaces of the laser cut sections. It is evident that the

Figure 8.10 a) Top and b) bottom surfaces of laser cut sections

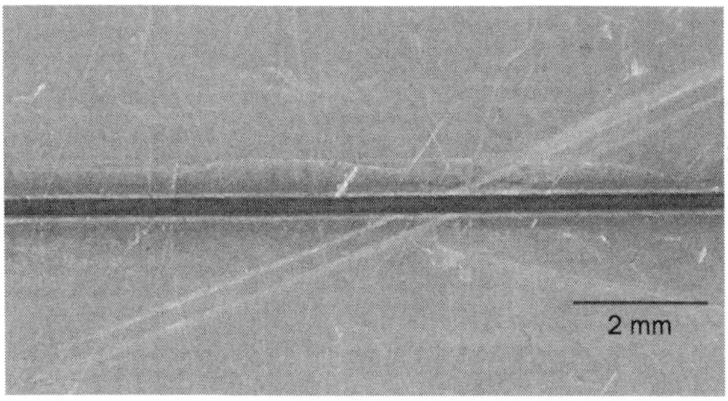

(a)

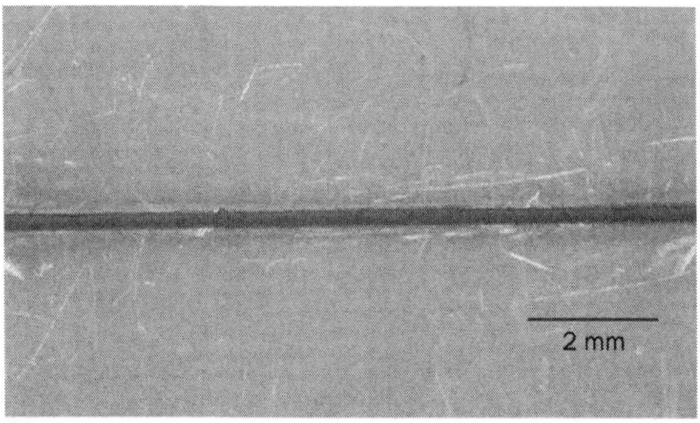

(b)

laser cut sections are parallel and free from excessive defects such as large area sideways burns. Figure 8.11 shows the close view of optical micrographs for the top and bottom surfaces of the cut sections. It can be observed that some

Figure 8.11 Close view of a) top and b) bottom surfaces of cut sections

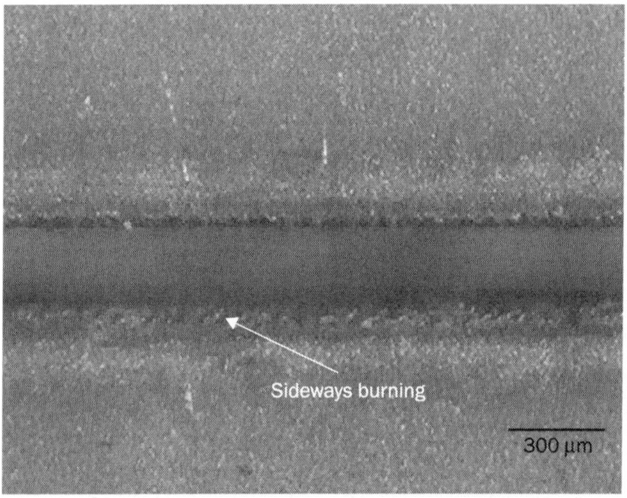

(a)

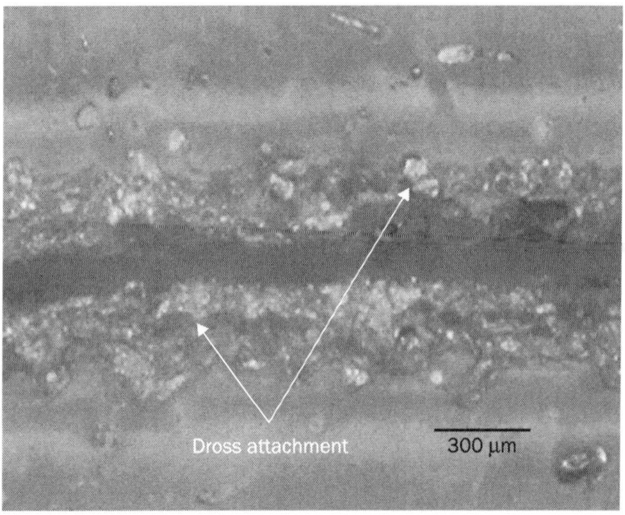

(b)

small sideways burning occurs around the kerf edges. This can be attributed to the presence of oxygen due to oxidation of the workpiece surface prior to laser cutting. In this case, the presence of oxygen in the cutting section initiates high temperature oxidation reactions during the cutting process while generating a small amount of sideways burning around the cut sections, despite the fact that nitrogen at high pressure is used as the assisting gas. At the bottom surface of the laser cut section, some small dross attachments are observed. This occurs because of the molten flow which cools down with the presence of the assisting gas upon ejection from the kerf. Although temperature remains high at the bottom surface of the workpiece at a location where the laser beam passes, in the region behind the laser spot, temperature reduces at the cut section. This increases the viscosity of the molten flow, and the drag force, generated in the kerf and acting onto the molten flow due to the assisting gas, is not sufficiently high, resulting in total ejection of the molten material from the kerf. Nevertheless, the dross size is small and it is randomly distributed along the cut edges at the bottom side of the workpiece.

Figure 8.12 shows SEM micrographs of the kerf surface and the cross section of the laser cut region. The striation patterns are formed at the surface; however, the depth of the striation patterns is shallow. This is attributed to the frequency of the laser output power (Table 8.2), which minimizes the striation formations (Yilbas, 2004). In addition, some locally scattered surface cracks are observed at the kerf surface due to the formation of a brittle recast layer. This is because of the high cooling rates at the kerf surface. In addition, the use of nitrogen as an assisting gas causes the formation of nitride compounds in the recast layer vicinity. This contributes to the brittle structure in the recast layer. Nevertheless, microcracks do not form a crack network

Figure 8.12 SEM micrographs of the kerf surfaces

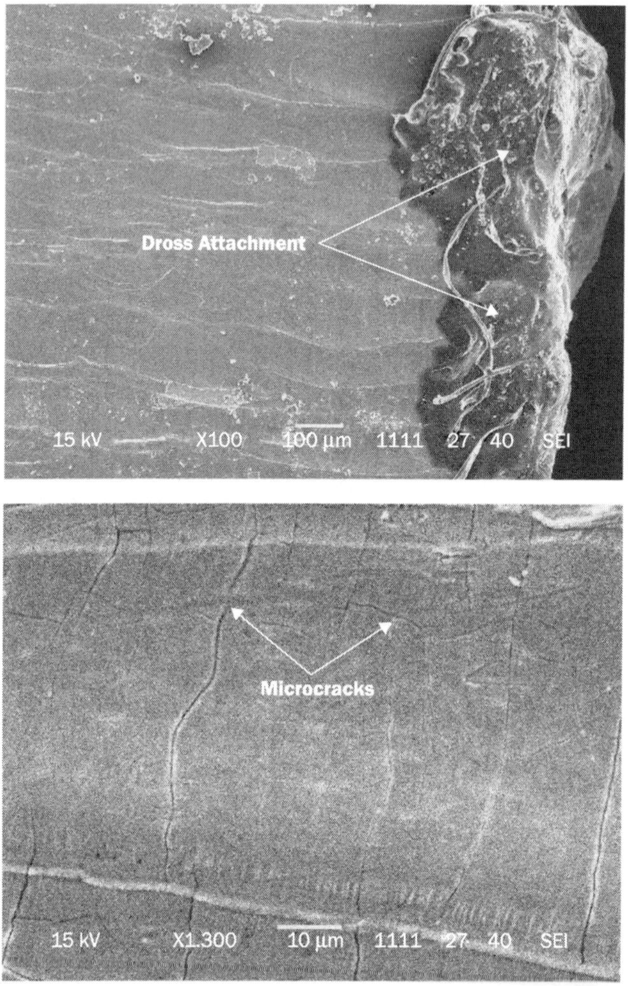

at the kerf surface. It is also evident from the cut region cross-sectional micrographs that the thickness of the recast layer is of the order of few micrometers and it is only formed at the kerf surface. The high cooling rates at the kerf surface cause development of the needle like structures behind the recast layer (Fig. 8.13). The heat affected zone is very shallow, which is attributed to the low thermal diffusivity of titanium alloy.

Figure 8.14 shows XRD diffractogram of the laser cut section. The presence of a Ti_2O peak reveals that the surface of the workpiece in the cutting section contains oxygen, which causes high temperature oxidation reactions during the cutting process while resulting in some small sideways burning. The use of high pressure nitrogen gas resulted in TiN_x formation in the cutting section. This contributes to the brittleness of the surface through increasing the surface hardness of the recast layer. When comparing the residual stress predicted (740 MPa) and measured from the XRD technique (720±30 MPa), it is evident that both results are in good agreement.

Figure 8.13 SEM micrograph of cross-section of laser cut section in the kerf vicinity

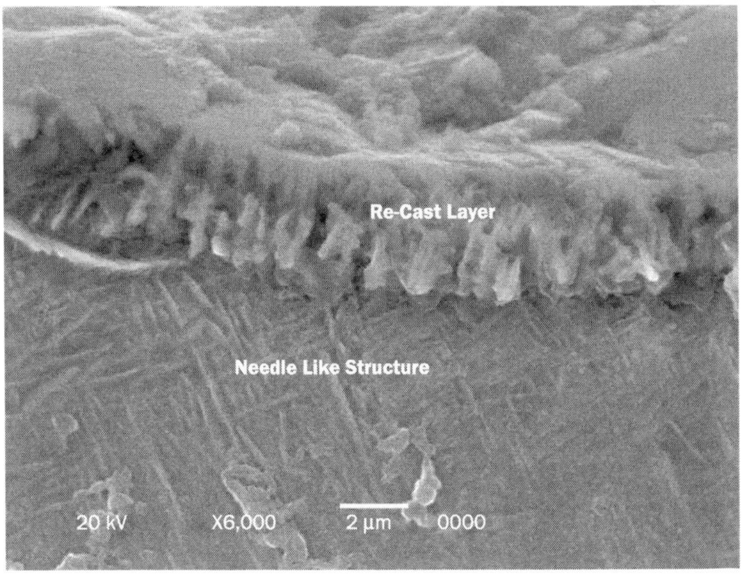

Figure 8.14 XRD diffractogram in the cutting region

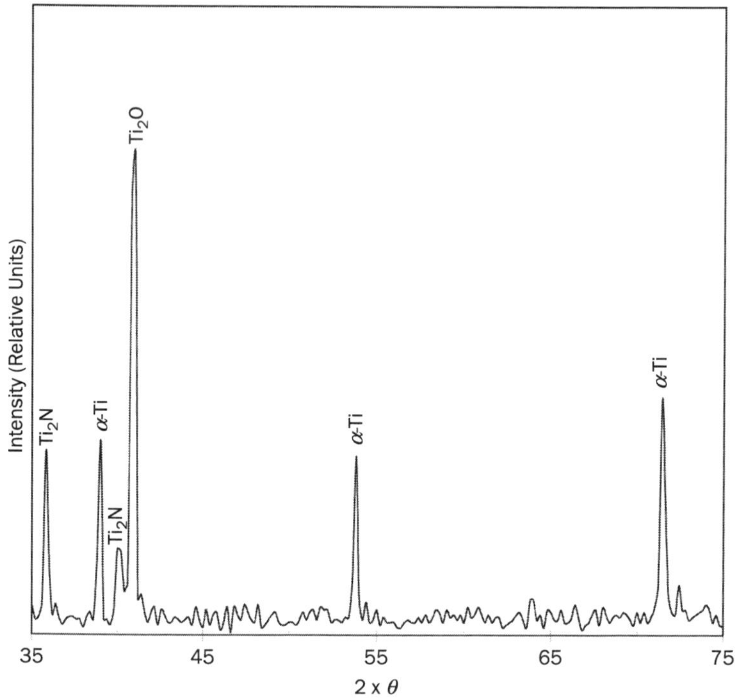

8.5 Conclusions

Laser straight cutting of titanium alloy has been carried out. Thermal stress field in the cutting section has been predicted using ABAQUS finite element code. The morphological and microstructural changes in the cutting section have been examined using optical and scanning electron microscopes, and *x*-ray diffraction. It was found that the temperature rises well above the melting temperature of the alloy in the laser irradiated region during the cutting process. This, in turn, results in superheating of the liquid phase and slowing down the cooling rate in this region. The temperature gradient remains high in the frontal region of the irradiated spot due to large differences in the temperatures between the irradiated spot and the base material. Von Mises stress remains low in the early cooling periods because of high temperature, which lowers the elastic modulus of the alloy. Von Mises stress attains high values where the temperature gradients are high. The maximum value of von Mises stress occurs in the vicinity of the top and bottom surfaces. Von Mises stress increases at the mid-thickness of the workpiece due to compressive stresses formed in this region.

The resulting cut section is parallel and some small sideways burns are observed at the top surface of the cut edges. This is associated with the presence of oxygen at the workpiece surface prior to laser cutting process, which is also observed from the *x*-ray diffractogram peaks. The dross attachment takes place at the bottom surface of the cut section, which is associated with increasing viscosity of the molten flow and low drag force generated by the assisting gas in the kerf. The use of nitrogen results in the formation of TiN_x in the cutting section, which contributes to increased brittleness of the recast layer. Randomly distributed microcracks are observed at the recast layer surface because of the brittle structure. The thickness of the recast layer is of the order of a few micrometers and a needle-like structure is formed in the region below the recast layer due to high cooling rates. The depth of heat affected zone is very shallow in the cut region because of the low thermal diffusivity of titanium alloy.

8.6 Acknowledgements

The authors acknowledge the support of King Fahd University of Petroleum and Minerals Dhahran, Saudi Arabia and Gazi University in Turkey for this work.

8.7 References

ABAQUS (2009), *ABAQUS Theory Manual Version 6.7*. Pawtucket: ABAQUS Inc.

Almeida I A, de Rossi W, Limab M S F, Berretta J R, Nogueira G E C, Wetter N U, Vieira Jr N D (2006), 'Optimization of titanium cutting by factorial analysis of the pulsed Nd: YAG laser parameters', *J. Materials Processing Technology*, **179**: 105–10.

Deng S, Matsunawa A, Lawrence Y, Zhong M (2005), 'Lasers in material processing and manufacturing II', *SPIE Proceedings*, **5629**: 247–52.

H&M Analytical Services, Inc., Available from: http://www.h-and-m-analytical.com/pdfs/residual_stress.pdf. [accessed 10 December, 2010].

Khana Z A, Hadfield M, Tobe S, Wang Y (2005), 'Ceramic rolling elements with ring crack defects – a residual stress approach', *Materials Science and Engineering A*, **404**: 221–6.

Maher W, Tong K (1998), 'Laser cutting of titanium with pulsed and modulated pulsed Nd:YAG lasers', *Proc. SPIE*, **3265**: 180–3.

Pfeifer R, Herzog D, Hustedt M, Barcikowski S (2010), 'Pulsed Nd:YAG laser cutting of NiTi shape memory alloys—Influence of process parameters', *J. Materials Processing Technology*, **210**: 1918–25.

Rao B T, Kaula R, Tiwarib P, Nath A K (2005), 'Inert gas cutting of titanium sheet with pulsed mode CO_2 laser', *Optics and Lasers in Engineering*, **43**: 1330–48.

Shanjin L, Yang W (2006), 'An investigation of pulsed laser cutting of titanium alloy sheet', *Optics and Lasers in Engineering*, **44**: 1067–77.

Shen Z H, Zhang S Y, Lu J, Ni X W (2001), 'Mathematical modeling of laser induced heating and melting in solids', *Optics and Laser Technology*, **33**: 533–7.

Thawari G, Sarin S J K, Sundararajan G, Joshi S V (2005), 'Influence of process parameters during pulsed Nd:YAG laser cutting of nickel-based superalloys', *J. Materials Processing Technology*, **170**: 229–39.

VERIDIAM, Available from: http://www.veridiam.com/pdf/DataSheetTitaniumAlloy.pdf [accessed 01 December, 2010].

Yang J, Sun S, Brandt M, Yang W (2010), 'Experimental investigation and 3D finite element prediction of the heat affected zone during laser assisted machining of Ti6Al4V alloy', *J. Materials Processing Technology*, **210**: 2215–22.

Yilbas B S (2004), 'Laser cutting quality assessment and thermal efficiency analysis', *J. Materials Processing Technology*, **155–6**: 2106–15.

Yilbas B S, Akhtar S S, Karatas C (2011), 'Laser trepanning of a small diameter hole in titanium alloy: Temperature and stress fields', *J. Materials Processing Technology*, **211**(7): 1296–304.

Index

$230.00